Radio images taken at intervals of one month show blobs of gas being ejected from SS433, an object thought to be a black hole or a neutron star with a unique pair of opposed, high-velocity jets. In the first frame *(far left)*, budlike appendages appear on each side of the object; in successive frames, they move farther from the center, finally detaching as new clumps of matter begin to push their way out of the nucleus.

for example, emit radiation, and interactions between electrically charged particles also produce electromagnetic waves *(pages 49-59)*. These waves have a peculiar dualistic character: They can also be described as particles, depending on the measuring instruments used. The particle aspect of radiation, called a photon, is a discrete bundle with a measurable energy; a wave is a continuous stream with a wavelength and a frequency.

All radiation carries information about the nature of the event that produced it. Any heated object, for example, emits radiation over a range of wavelengths, with a concentration around a single wavelength. A very hot object produces predominantly short waves, while a cooler body radiates longer ones. Thus a piece of iron placed in a forge begins to glow with the long waves of low-frequency red light, and as it heats up, it gives off shorter-wavelength yellow light. Similarly, clouds of gas only a few degrees warmer than absolute zero radiate long infrared and radio waves; the hottest stars give off short-wavelength x-rays and gamma rays.

Many other processes contribute to the radiation coursing through the universe. Every element emits and absorbs energy according to a characteristic pattern, which astronomers use to identify the chemical composition of distant objects. Electrically charged particles, interacting with each other or with a magnetic field, produce electromagnetic waves with certain known attributes; one of the most common is synchrotron radiation, emitted by electrons spiraling in a magnetic field. The human view of these remote activities, though, has been severely restricted. The upper layers of Earth's atmosphere absorb all but a small fraction of celestial radiation, and eyes can see only part of what gets through.

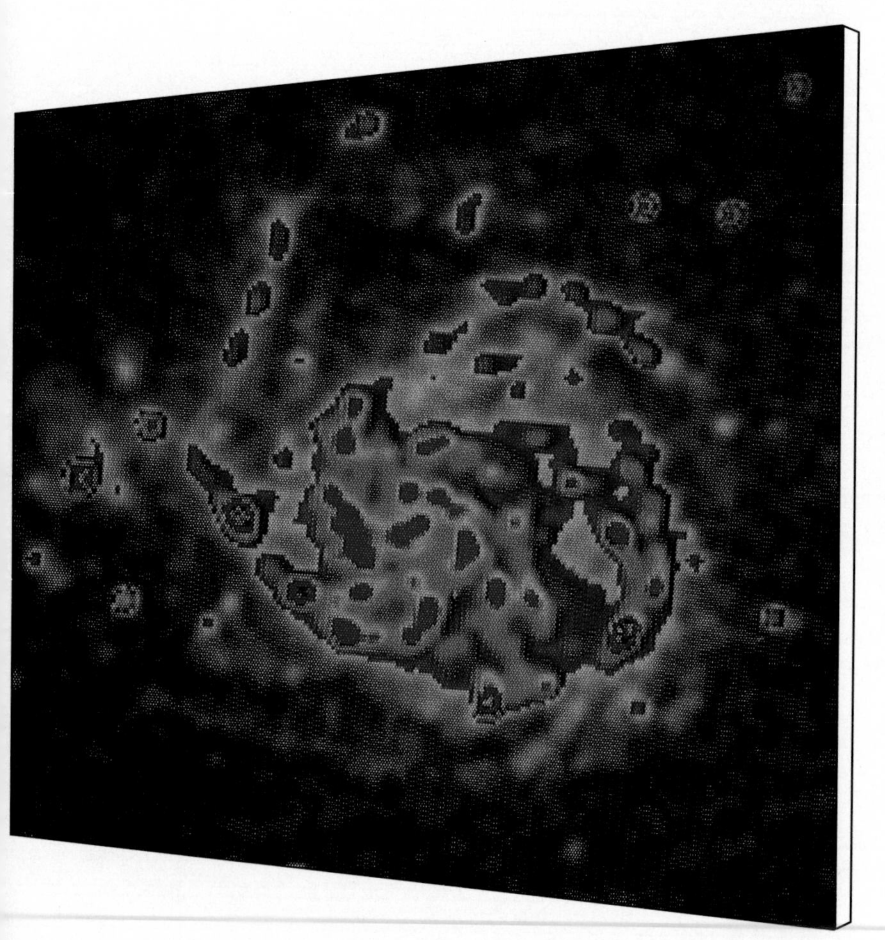

For thousands of years, the unaided eye was the sole astronomical instrument. It was well adapted to the visible light of the Sun, by far the most important source of radiation that reaches Earth. Humans who studied the sky saw what there was to see: the brightest stars, like the Sun, shining mostly at optical wavelengths, and the Moon and planets, which reflect the Sun's light. Intrinsically dim objects, as well as distant ones and those that emit only nonoptical radiation, remained invisible.

In 1610, the picture of the heavens was changed forever when Italian astronomer Galileo Galilei looked through a crude telescope, discovering that the cloudlike Milky Way was in fact "a congeries of innumera-

ble stars." Galileo went on to make hundreds of other discoveries, including mountains on the Moon and four satellites around Jupiter.

Galileo's telescope, like all subsequent collectors of electromagnetic radiation, operated on a simple principle. Its glass lenses bent incoming light waves, concentrating and focusing them at an eyepiece for the observer to see. The telescope had two advantages over even the most acute human eye: The magnifying power of its lenses gave it superior resolution (the ability to discern small details), and its three-inch-wide tube, with about fifty times more light-gathering area than the pupil of an eye, gave it greater sensitivity (the ability to make out faint objects). Resolution transformed the haze of the Milky Way into individual stars; sensitivity allowed the discovery of the faint Jovian moons.

Succeeding generations of astronomers greatly improved on Galileo's instrument. They built larger telescopes, which captured more light for better sensitivity, and designed bigger and more-sophisticated optics—compound lenses at first, and then curved mirrors as well—to enhance resolution. By the middle of the nineteenth century, scientists began the regular use of a technique called spectroscopy to analyze the content of the light itself. With a prism (or later, other optical devices), a researcher could break the light of the Sun or a star into a colored spectrum, then measure the intensity of the

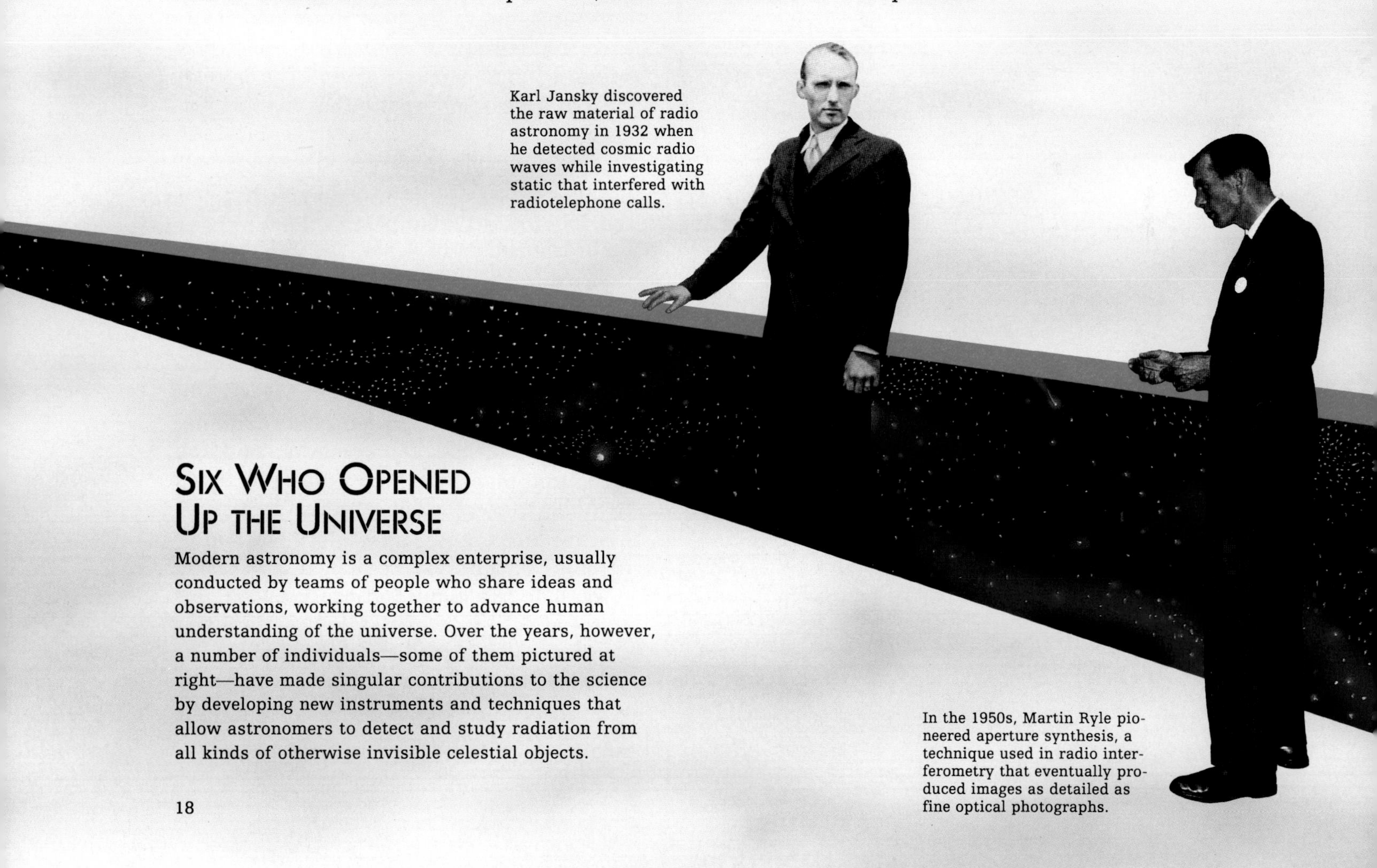

Karl Jansky discovered the raw material of radio astronomy in 1932 when he detected cosmic radio waves while investigating static that interfered with radiotelephone calls.

Six Who Opened Up the Universe

Modern astronomy is a complex enterprise, usually conducted by teams of people who share ideas and observations, working together to advance human understanding of the universe. Over the years, however, a number of individuals—some of them pictured at right—have made singular contributions to the science by developing new instruments and techniques that allow astronomers to detect and study radiation from all kinds of otherwise invisible celestial objects.

In the 1950s, Martin Ryle pioneered aperture synthesis, a technique used in radio interferometry that eventually produced images as detailed as fine optical photographs.

different colors, which correspond to distinct ranges of wavelengths. Comparing the characteristics of the light to the results of laboratory experiments, astronomers could decipher the chemical composition of celestial bodies that emit or absorb light. With every new development, vision through the optical window in the atmosphere became clearer.

The first sign of another window was discovered by the British astronomer Sir William Herschel in 1800. Performing a crude analysis of sunlight with a prism, he was surprised when a thermometer left in the apparently dark area just beyond the red end of the spectrum registered a substantial temperature rise. Subsequent investigation showed that this band of radiation was relatively narrow; the heat dwindled to nothing a short distance from the visible spectrum.

Infrared radiation (so called for its position just beyond visible red light) was still an astronomical curiosity in 1881, when the American scientist Samuel Pierpont Langley developed an instrument called a bolometer, an electrical detector of radiant heat over a broad range of wavelengths. From a vantage point high up Mount Whitney in the Sierra Nevadas of California, above much of the atmospheric blanket, Langley measured the Sun's energy through the visible section of the spectrum and then beyond—and found that the bolometer registered heat at much longer wavelengths, far past the region of Herschel's discovery.

Although neither fully realized the significance of his explorations, Herschel and Langley had begun to chart much of the infrared wave band.

Frank Low developed a highly sensitive infrared detector in the 1960s, paving the way for the discovery of many objects not quite hot enough to radiate visible light.

In 1970, Robert Wilson *(far left)* and Arno Penzias used a millimeter-wave receiver to detect celestial carbon monoxide, an important molecule in vast interstellar clouds.

Lyman Spitzer, Jr., an early champion of astronomical satellites, led the team that developed an ultraviolet telescope launched into Earth orbit in 1972.

Herschel's discovery, now known as the near infrared band, includes radiation with wavelengths of about one micron (a millionth of a meter); near infrared is commonly emitted by matter heated to about 2,000 degrees Kelvin. (The Kelvin scale uses Celsius degrees but sets 0 degrees at absolute zero, or −273 degrees centigrade.) The middle infrared, which Langley found, with wavelengths as long as ten microns, is produced by cooler bodies, with temperatures down to about 500 degrees.

The combined findings of Herschel and Langley made it clear that infrared waves were an important component of solar energy. Astronomers assumed that stars, generally believed to be similar to the Sun, must also be infrared sources. Langley's bolometer, however, was not sensitive enough to detect the trifling amount of this radiation completing the journey from a distant star; half a century passed before such measurements were possible, and then only crudely. By that time, radio, the next form of invisible radiation to be discovered in the heavens, was moving to center stage.

STATIC FROM SAGITTARIUS

Astronomical observations at radio wavelengths started by accident, in a study performed in the early 1930s by Karl Jansky, an American communications engineer. Working for Bell Telephone Laboratories in New Jersey, Jansky was assigned to track down the sources of static that interfered with ship-to-shore radiotelephone communications. Among the signals Jansky detected was one that clearly originated in the heavens, moving with the stars and apparently centered somewhere in the region of the constellation Sagittarius—the direction of the center of the Milky Way galaxy.

The discovery created a mild stir in the scientific community, but no established astronomers were moved to abandon their study of visible objects to follow Jansky's lead. For some years, the only research in radio astronomy was conducted by Grote Reber, a young American electronics engineer who assembled a thirty-one-foot parabolic dish antenna in his backyard in a Chicago suburb. Although the device was primitive, it embodied the principles that would govern the design of every subsequent radio telescope. Radio waves striking the metal strips that made up the dish were reflected to a radio receiver at the focus of the parabola. Reber began recording radio signals from the sky in 1937, reporting his findings from time to time in professional journals for radio engineers and astrophysicists. His work culminated in a map of radio sources scattered through the Milky Way, finished in 1944.

For all his assiduous cataloging, Reber could only speculate about the origins of the signals. By the mid-1940s, however, a few astrophysicists were developing theories about physical processes that might have distinctive signatures at radio wavelengths. One was a Dutch scientist named Hendrik van de Hulst, who in 1944 predicted a characteristic of hydrogen atoms that would cause clouds of the gas to radiate at a wavelength just over twenty-one centimeters, in the radio band. Such clouds, which would not ordinarily emit visible light, had been discovered by indirect means decades earlier: Ob-

Debris from a star that exploded 300 years ago in the constellation Cassiopeia radiates at various wavelengths in this composite image of the supernova remnant, Cassiopeia A. Very hot gas emits x-rays *(green)*, visible light *(red)* emanates from cooler gas clouds and radio waves *(blue)* come from particles in magnetic fields

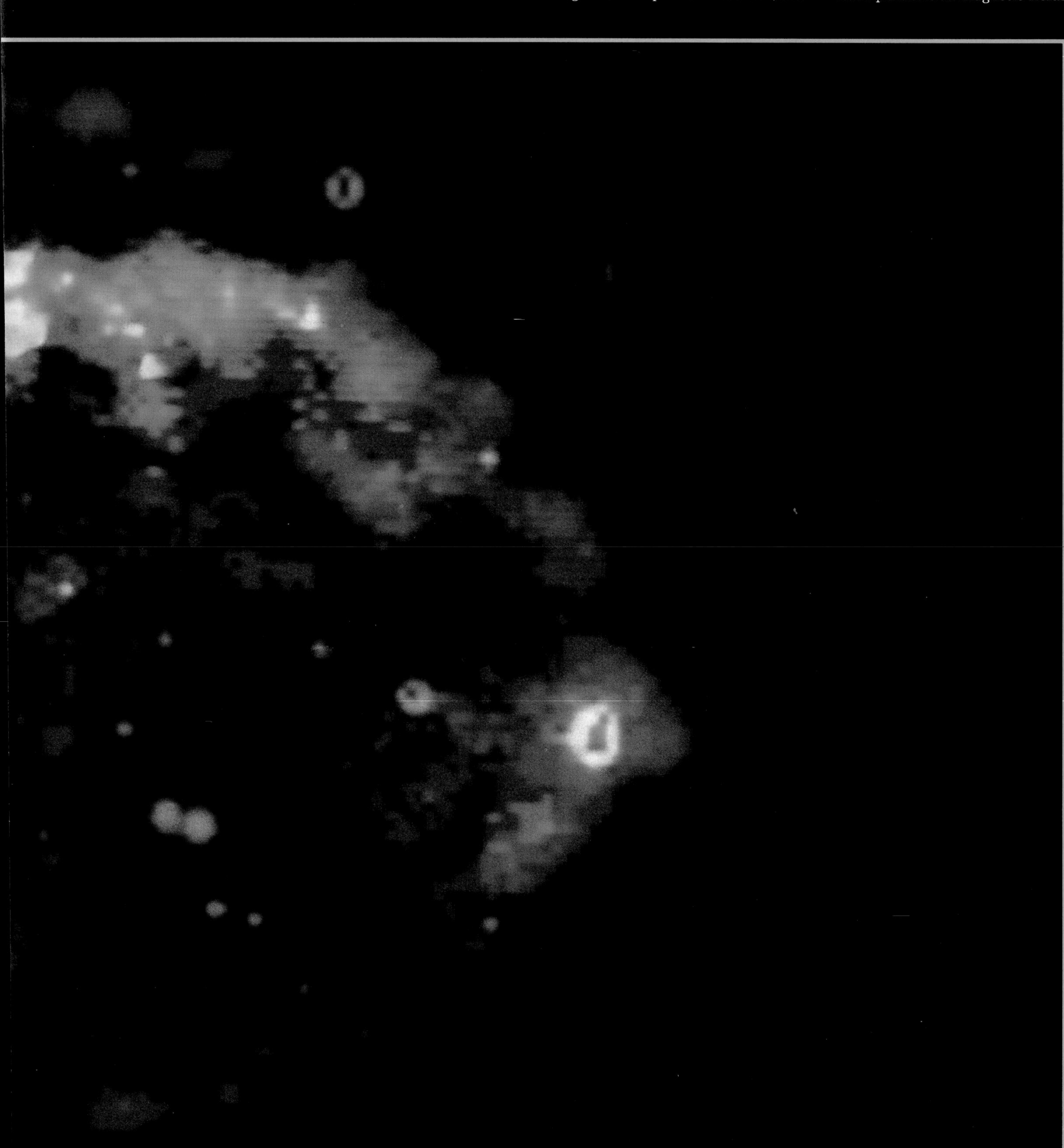

Deep in the Milky Way, two brilliant beams pierce trillions of miles of space, illuminating the tenuous gas between the stars like flashlights in a dark, smoke-filled room. Although they glow with visible light and x-rays, these are not light beams; they are actually tightly focused jets of hydrogen and helium, moving nearly 50,000 miles per second—roughly a quarter the speed of light. The radiance emanates from molecules of the interstellar gas, heated by friction to temperatures hundreds of times greater than on the surface of the Sun. Floods of radiation outline the entire lengths of the jets, which dissipate in giant puffs of disturbed gas 100 light-years from their origin.

The source of the jets is an unlikely pair of stars, twirling through a bizarre celestial waltz. One dancer is a normal star, a hot, bluish ball of hydrogen about twenty times the size of the Sun, blazing with the heat of internal nuclear reactions. Its partner, a cold sphere that emits no light, is only ten miles across—yet is more massive than the Sun. The tiny ball is the collapsed remnant of a star, packed so densely that a thimbleful weighs as much as 2,000 battleships.

Locked in a tight gravitational embrace, the two stars circle each other at a range of only 20 million miles, one-fifth the distance between Earth and the Sun. The powerful gravity of the dense collapsed star rips matter from the surface of its companion, sucking a continuous stream of hydrogen and helium across the gap. Prevented by its momentum from falling directly onto the collapsed star, the gas swirls instead into a disk around it. The sheer volume of falling gas—as much as a million billion tons every second—builds up enormous pressure in the disk. At irregular intervals, the squeeze becomes too tight, and blobs of gas squirt out, forming the two opposed jets perpendicular to the disk.

Torn free of their atoms in the maelstrom of the jets, swirling electrons emit radio waves with a combined power unimaginably greater than that of all the radio stations on Earth. Together with the visible light and x-rays, the radio waves spread through space, their intensity dwindling with every mile they traverse. When the radiation reaches Earth, 15,000 light-years away, it is diminished to almost imperceptible whispers. Nevertheless, the light, concentrated by the curved mirror of a giant telescope, yields its secrets to a variety of electronic devices; the x-rays trigger detectors on a satellite or-

biting Earth at an altitude of several hundred miles; and the radio waves, intercepted by dozens of linked antennas, wend their way through amplifiers and computers to form colorful images of the jets on a video screen.

Billowing across the spiral arm of a galaxy, a vast cloud of dust and gas harbors a stellar nursery, where stars slowly coalesce and flare into life. Nascent stars in all stages of development coexist in the hundred-light-year expanse of the cloud. In some regions, a slight thickening of matter—still far more rarefied than the lightest terrestrial mist—heralds the earliest evolutionary stage. Elsewhere, similar clumps are shrinking under the force of their own gravity, to consolidate during millions of years into loose spheres called protostars. The interior of the cloud is already speckled with protostars, incandescent with the heat generated by their continuing gravitational collapse. They are interspersed with the blazing orbs of stars that have completed their gestation, achieving the high density and million-degree temperatures required to ignite their nuclear fires. The fierce radiation from the hot new stars pushes out through the cloud, compressing the gas in nearby areas and triggering new rounds of star formation.

The immense cloud absorbs most of the high-energy stellar radiation, leaving only a trickle of ultraviolet waves to fan out from the galaxy. But what the cloud absorbs, it gives back in a different form. Dust particles, warmed by the energy they take in, act like tiny embers, releasing their heat into space as infrared rays. When the faint traces of this radiation reach Earth, millions of light-years away, the terrestrial atmosphere blocks its passage. Human curiosity and ingenuity have surmounted this vaporous barrier, however. Far above the obscuring gases, a satellite transforms the infrared energy into images that show the great expanse of the cloud. And a rocket-borne detector, briefly arching seventy-five miles above Earth's surface, captures ultraviolet snapshots that show the stars themselves, fiery masses that make the galaxy one of the brightest in the heavens.

These wonders and many more are the regular fare of today's astronomers, who watch the cosmos with eyes not limited by the narrow scope of human vision. Their instruments give them access to the entire spectrum of electromagnetic radiation, which includes such phenomena as radio and infrared waves, visible and ultraviolet light, and x-rays. Made up of waves of energy that travel at the speed of light, electromagnetic radiation is the primary medium by which remote cosmic events are known.

Scientists sift volumes of information from the electromagnetic murmurs of distant stars and galaxies, aided by devices that can pinpoint sources of invisible radiation with uncanny accuracy. Their tools are fruits of a technology of discovery that has transformed astronomy since the middle of the twentieth century. In just a few revolutionary decades, astronomers have moved from almost total reliance on optical telescopes to the routine use of continent-spanning arrays of radio telescopes and delicate satellite-borne

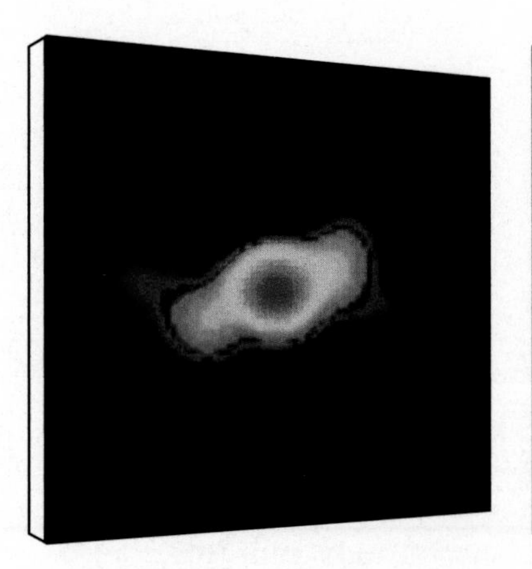
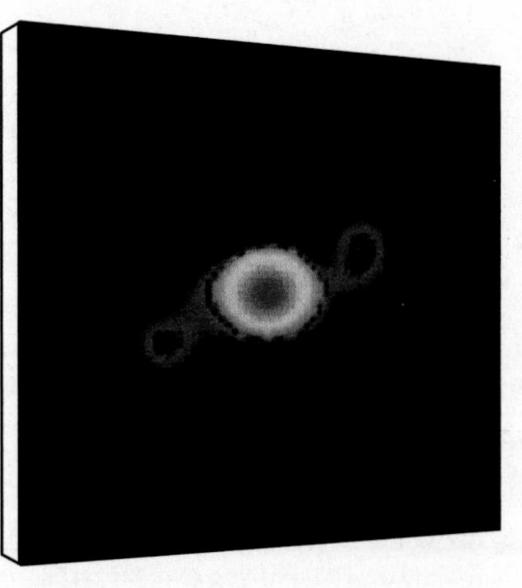
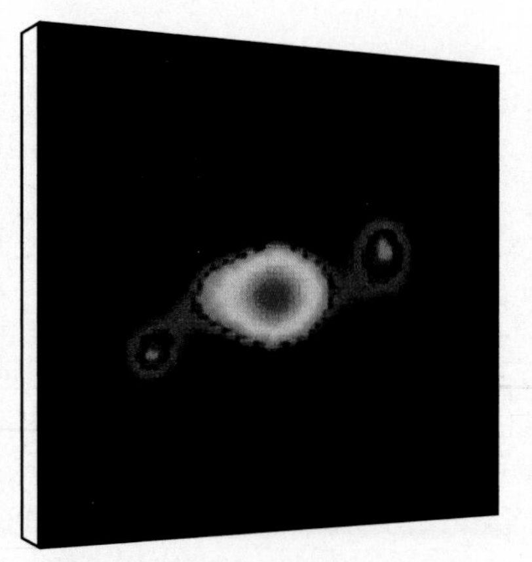
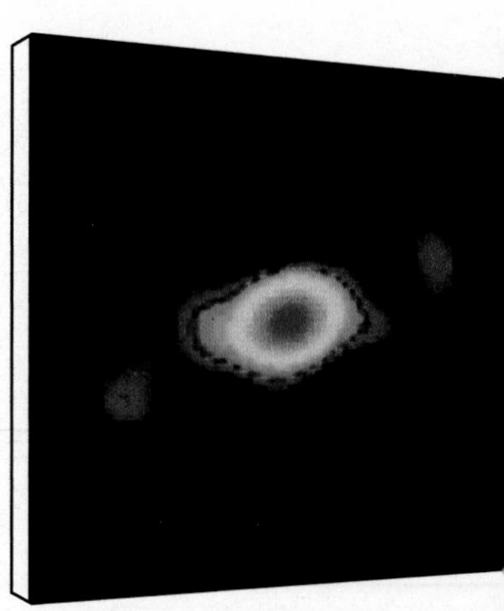

detectors. Today, scientists can view the universe with nearly equal sensitivity across most of the electromagnetic spectrum, achieving a level of detail that would clearly show a penny more than three miles away. They can also decipher messages carried by nonelectromagnetic celestial couriers such as cosmic rays and weightless neutrinos.

Conceptual breakthroughs have paralleled the technological upheaval. In the 1940s, astronomical theory was geared to interpreting the gross workings of the cosmos. Relationships between stars in a galaxy, for example, were well explained by the mathematical formulas of celestial mechanics, which describe the trajectories and orbits of bodies on the basis of their gravitational interactions. But the science of astrophysics, which involves studying the physical nature of the objects themselves, was still young. Although scientists had learned to analyze light for information about the composition of such objects as stars and gas clouds, there was little understanding of interactions at a small scale, including the processes that produced the light in the first place. As the century passed its midpoint, however, astrophysicists began to employ analytical tools devised by experimental chemists and nuclear physicists, learning to account for cosmic phenomena at the level of atoms and molecules. They also began turning to computers, both to reduce the drudgery of data analysis and to apply entirely new techniques, such as the simulation and testing of hypotheses, that enriched their picture of the universe with ever-greater depth and detail.

Rich in new stars, the galaxy M101 flares with ultraviolet radiation in this image from a rocket-borne telescope. Prominent concentrations of young stars show as red and yellow regions along the inner arms. Blue areas reveal thin clouds of dust and gas in the outermost parts of the galaxy. Almost invisible at optical wavelengths, the clouds scatter ultraviolet radiation emanating from stars that are closer to the center of the galaxy.

REVELATIONS

The new astronomy resulting from the technological and theoretical leaps has opened windows on a universe teeming with violent and extraordinary activities. Scientists now watch galaxies forming at the edge of the universe, ablaze with radiation from newborn stars and thick billows of hot gas. They probe the tattered ruins of long-dead massive stars, search for planets circling distant suns, and listen to the faint reverberations of the Big Bang—the primal burst of energy that gave birth to the cosmos. Revelations have come in a flood, with no end in sight.

Fortunately for the cause of astronomical discovery, electromagnetic radiation is generated by many different physical processes. All heated bodies,

1/An Expanding Spectrum

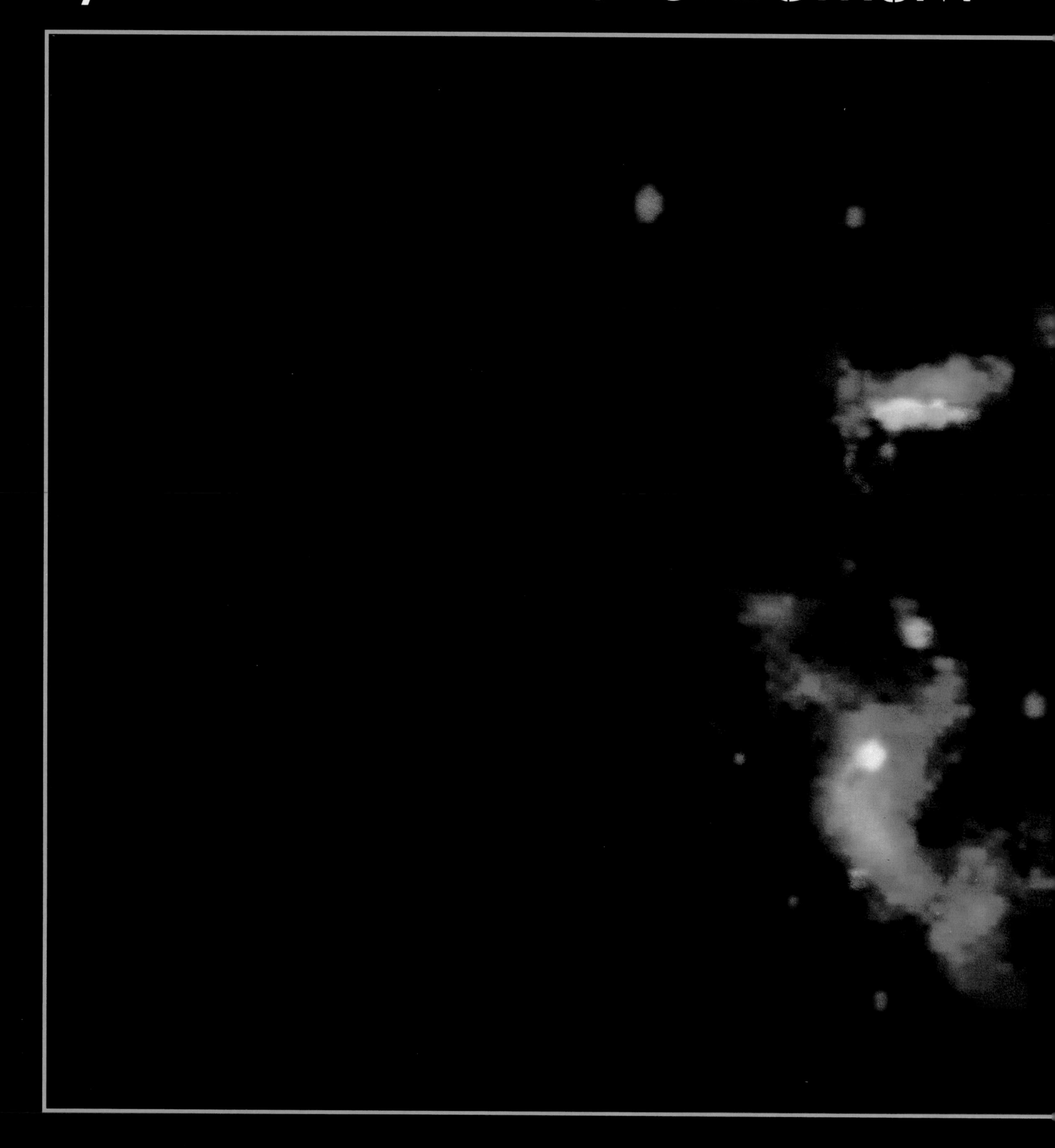

CONTENTS

VOYAGE THROUGH THE UNIVERSE

THE NEW ASTRONOMY

BY THE EDITORS OF TIME-LIFE BOOKS
ALEXANDRIA, VIRGINIA

Other Publications:
AMERICAN COUNTRY
THE THIRD REICH
THE TIME-LIFE GARDENER'S GUIDE
MYSTERIES OF THE UNKNOWN
TIME FRAME
FIX IT YOURSELF
FITNESS, HEALTH & NUTRITION
SUCCESSFUL PARENTING
HEALTHY HOME COOKING
UNDERSTANDING COMPUTERS
LIBRARY OF NATIONS
THE ENCHANTED WORLD
THE KODAK LIBRARY OF CREATIVE PHOTOGRAPHY
GREAT MEALS IN MINUTES
THE CIVIL WAR
PLANET EARTH
COLLECTOR'S LIBRARY OF THE CIVIL WAR
THE EPIC OF FLIGHT
THE GOOD COOK
WORLD WAR II
HOME REPAIR AND IMPROVEMENT
THE OLD WEST

This volume is one of a series that examines the universe in all its aspects, from its beginnings in the Big Bang to the promise of space exploration.

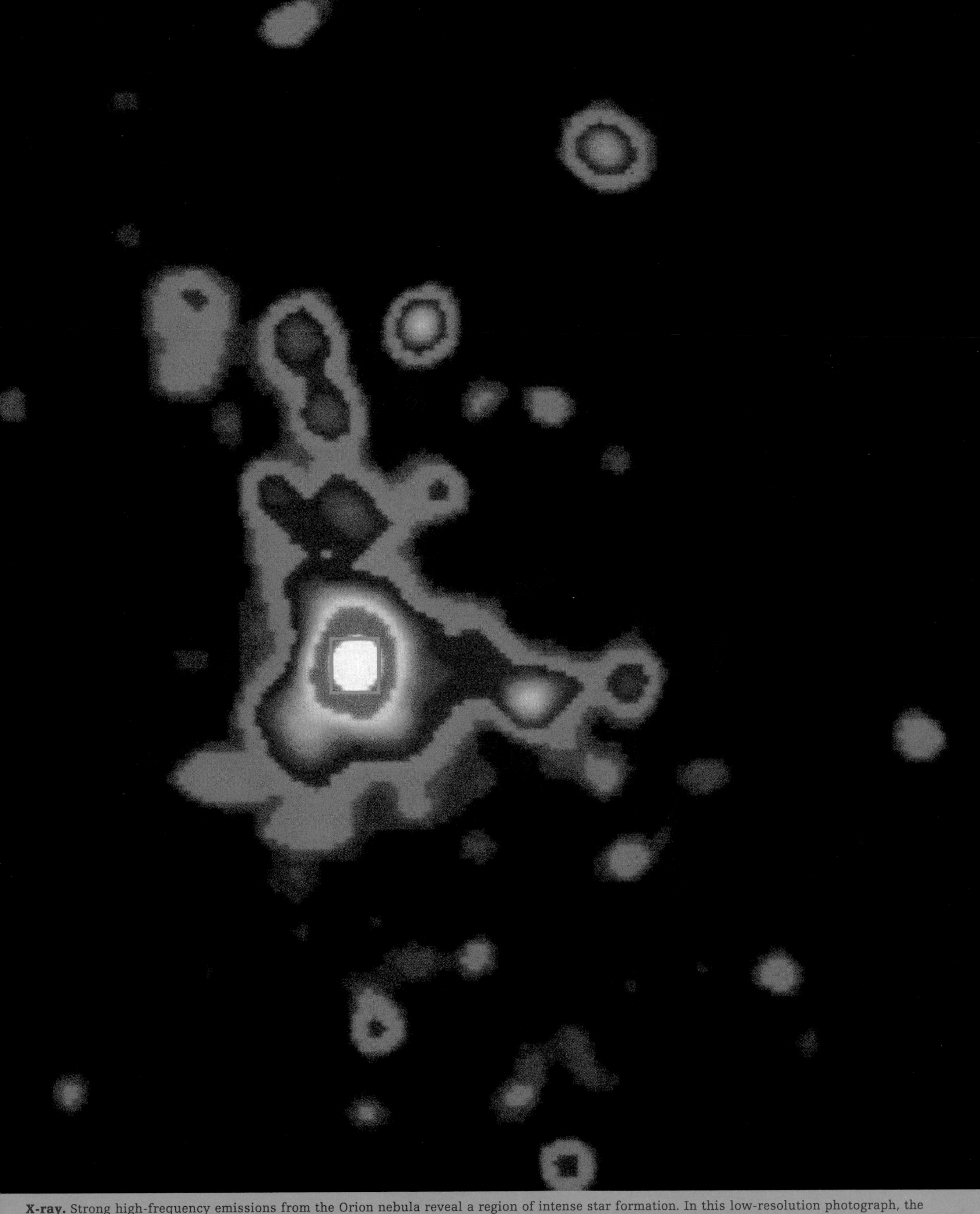

X-ray. Strong high-frequency emissions from the Orion nebula reveal a region of intense star formation. In this low-resolution photograph, the nebula's white core radiates as one huge x-ray source, powered mainly by massive newborn stars in and around the Trapezium cluster.

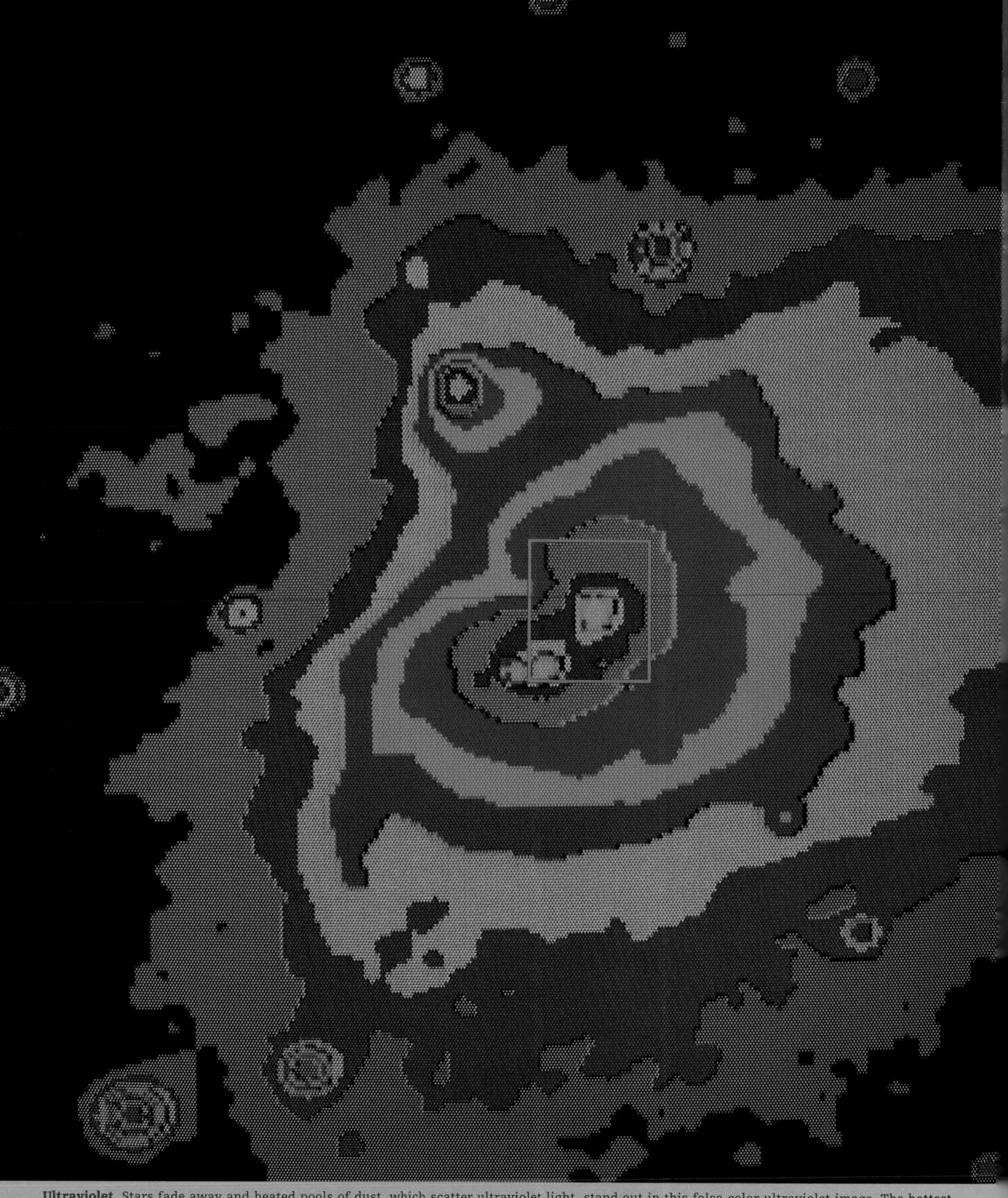

Ultraviolet. Stars fade away and heated pools of dust, which scatter ultraviolet light, stand out in this false-color ultraviolet image. The hottest areas of stellar debris shine blue around the central nebula and its companion, M43.

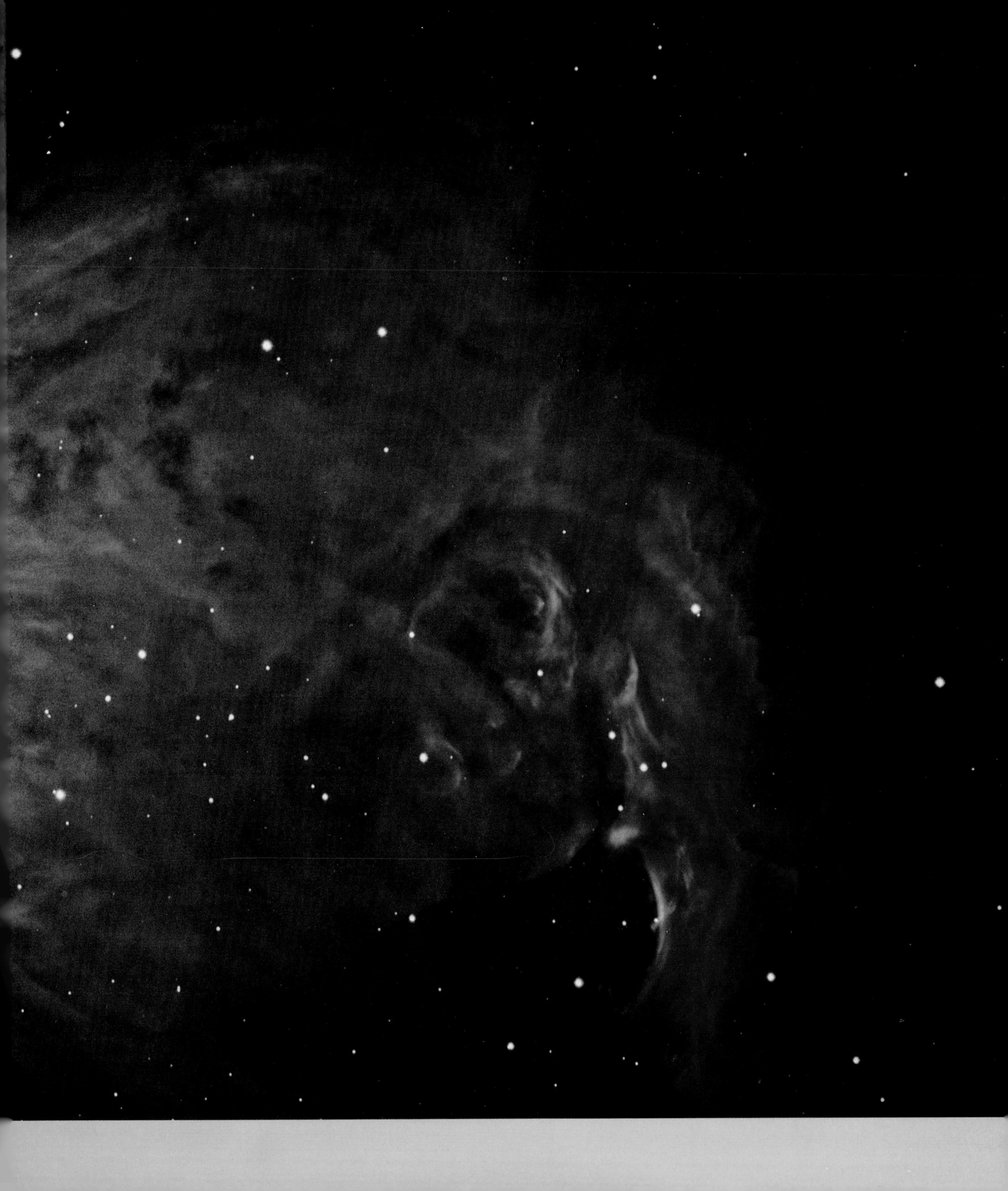

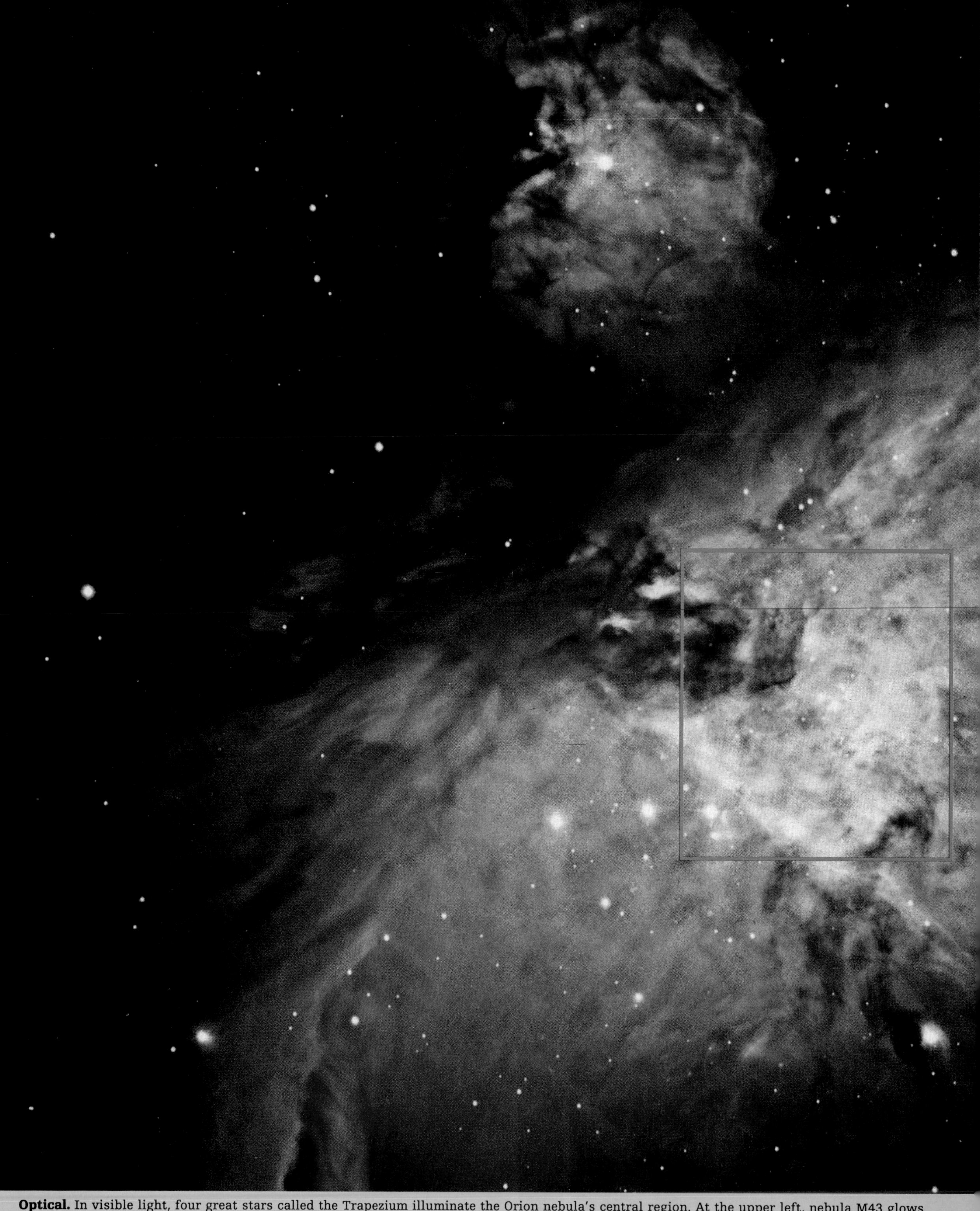

Optical. In visible light, four great stars called the Trapezium illuminate the Orion nebula's central region. At the upper left, nebula M43 glows around Nu Orionis, a star as bright as 24,000 Suns. (Here and on the following pages, the boxed area corresponds to the area on pages 2-3.)

Infrared. Just outside the visible spectrum, infrared light pierces the dust shrouding the Orion nebula to reveal a dense cluster of some 500 young stars. The area is a stellar nursery, cradling new stars in various stages of development; blue-white stars in the center are the hottest.

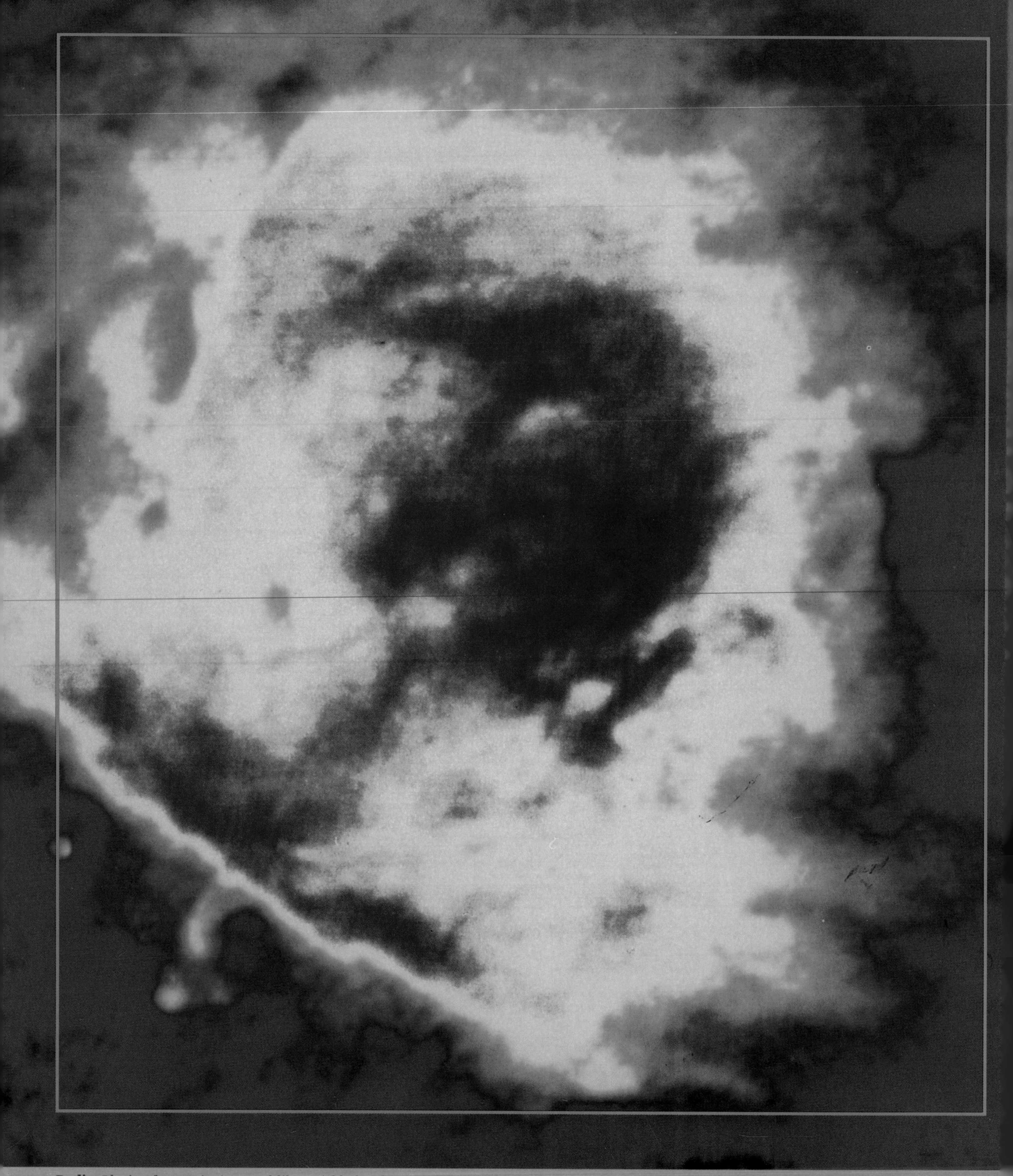

Radio. Blazing from Orion's sword like a richly hued gem, nebula M42 changes appearance—and yields up secrets—when viewed at wavelengths from radio to x-ray. Above, ionized gas in the nebula's center, heated to 14,000 degrees Fahrenheit, emits radio waves. Red areas are most intense.

THE NEW ASTRONOMY

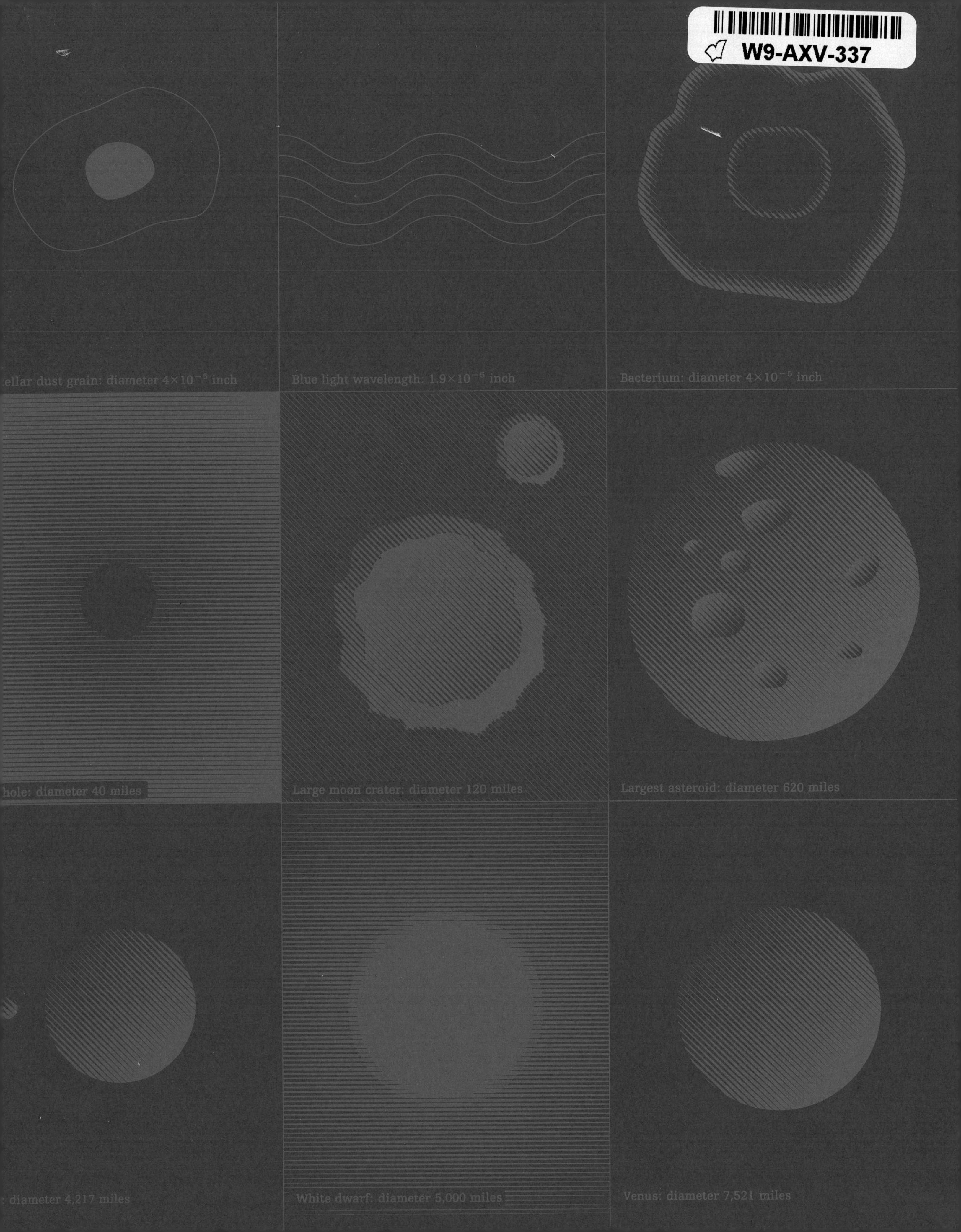
ellar dust grain: diameter 4×10^{-5} inch
Blue light wavelength: 1.9×10^{-5} inch
Bacterium: diameter 4×10^{-5} inch
hole: diameter 40 miles
Large moon crater: diameter 120 miles
Largest asteroid: diameter 620 miles
: diameter 4,217 miles
White dwarf: diameter 5,000 miles
Venus: diameter 7,521 miles

servers found that the light of some stars in the Milky Way contained dark absorption lines, evidence of selective absorption by rarefied gases in interstellar space. The absorption of starlight, however, only hinted at the extent of the gas clouds; no systematic study was possible because much of the galaxy is obscured by interstellar dust, impenetrable by visible light. Since radio waves can pass through dust, detection of the twenty-one-centimeter line would establish a means for surveying the clouds and examining otherwise invisible parts of the galaxy.

With World War II still raging in Europe, and his country under Nazi occupation, van de Hulst was unable to assemble the radio equipment required to find the twenty-one-centimeter line. When the conflict ended, however, the dismantling of the war machines on both sides released resources for radio astronomy. Many scientists and technicians had worked during the war to develop sensitive radar antennas and receivers, used to detect enemy ships and aircraft. In the postwar years, some turned their skills—and the now-surplus equipment—to the peaceful study of the heavens.

During a five-month span in 1951, researchers in the United States, the Netherlands, and Australia detected radio emissions from galactic hydrogen. The Dutch and Australian teams went on to chart the distribution of the gas clouds. Their final product, published in 1958 and known as the Leiden-Sydney map, confirmed what many astronomers had long believed: The Milky Way, like many other galaxies, has a spiral form, with extended arms delineated by vast fields of hydrogen.

RADIO SURPRISES

The exploitation of twenty-one-centimeter radiation was a kind of bridge between the old and new astronomies. An entirely new technology had been employed to prove a longstanding hypothesis about a classic question, the structure of the Milky Way. But for all the inventive theories and ingenious devices that made the breakthrough possible, the mapping of the galaxy held few surprises. Soon, however, radio astronomers began to stumble across totally unexpected phenomena.

One of the first and biggest surprises was a blue, starlike object in the constellation Virgo, invisible to the naked eye but discernible through small telescopes of the kind typically used by amateur astronomers. It was first observed, however, at radio wavelengths in the early 1960s, during a survey compiled by radio astronomers at Cambridge University in England. (Its name, 3C273, refers to its place in the resulting catalog.) Among the 471 radio sources listed, 3C273 ranks as the twentieth brightest. Astronomers, curious about the origins of these celestial radio signals, regularly sought to match them with visible objects. In a few cases they succeeded, identifying radio sources with dying stars or unusual galaxies, but the radio telescopes of the time could not pin down locations with enough precision to allow visual identification of the vast majority.

In 1962, Cyril Hazard, a British radio astronomer, came up with a novel

Ruining the View from Earth

For most people in the urbanized world, the star-thronged lanes of the Milky Way are becoming a rare sight, visible only on trips to the remote countryside. For astronomers, the situation is far worse: Increases in artificial lighting, radio-wave interference, space debris, and even radiation from nuclear-powered satellites have seriously degraded their view of the heavens. The photograph of city lights and agricultural fires at right—a mosaic of satellite images—illustrates the magnitude of the problem confronting optical astronomers. With the exception of the aurora borealis *(top left),* the bright patches outlining Earth's continents constitute light pollution.

Increased urban lighting has complicated astronomy in almost all of the United States, much of western Europe, and Japan. Optical astronomers typically observe objects many trillions of miles distant, sometimes collecting the light photon by photon over a period of hours. This process requires dark background skies. Not only do electric lights interfere, but city asphalt—an efficient reflector of light—intensifies the glow as well. Furthermore, automobiles and factories pump out emissions loaded with particulates that scatter light back toward the telescopes.

Nor is light the only problem. The cities that radiate visible light are also immersed in radio waves from television transmitters, cordless phones, and many other electronic sources. Satellites add to the electronic noise, beaming down signals with millions of times the power of cosmic sources. In space, nuclear-powered satellites emit high-energy particles that create false readings in sensitive gamma ray detectors. Finally, satellites and orbiting debris from thirty years of rocket launchings often leave light trails in astronomical photographs, obscuring faint stars. Moving faster than a bullet, the debris threatens planned space-based observatories.

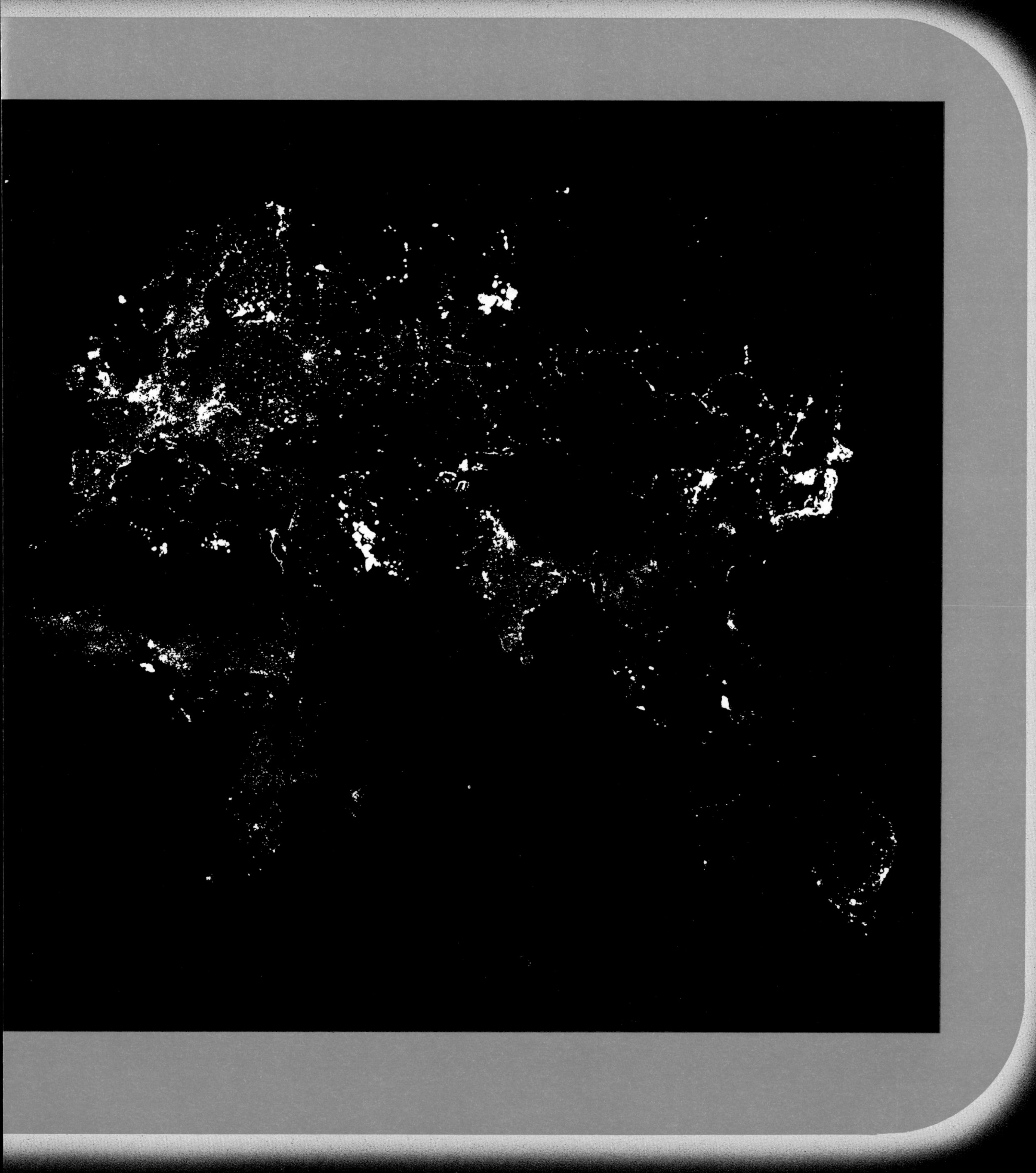

VANISHING STARS, ORBITING DEBRIS

Light pollution blazes out from a wide variety of sources, most notably street lamps, billboards, sports fields, and car dealerships. Scattered by atmospheric molecules, the glow can become so intense that it overwhelms starlight. At Mount Wilson Observatory near Los Angeles, California, the eighty-year-old 100-inch Hooker telescope has lost much of its effectiveness due to light and air pollution. Even at Kitt Peak National Observatory, located in the desert sixty miles outside of Tucson, Arizona, the nearby city's lights—though tightly regulated—make the night sky 6.5 percent brighter than it would be naturally. High-pressure sodium street lamps give off broad-spectrum light that scientists have difficulty filtering out. For this reason, astronomers strongly prefer that cities use narrow-spectrum, low-pressure sodium lighting, shielded to prevent scattering.

Meanwhile, the skies have become increasingly cluttered with inactive satellites and satellite fragments, spent rocket stages, and discarded equipment. While space debris has marred many astronomical photographs with light streaks, a greater threat from the 48,000 pieces of space junk is collision with new observatories going into orbit. For instance, the thirty-year accumulation of orbital flotsam has forced designers to add shielding to the Gamma Ray Observatory, lifting off in the early 1990s, and computers on the Hubble Space Telescope, due to go up in 1990, have been altered so they will not confuse the glittering junk with stars used for navigation.

The lights of Tucson, Arizona, were too dim in 1959 to pose a problem for Kitt Peak astronomers sixty miles away.

By 1980, Tucson's lights threatened Kitt Peak's telescopes. Strict lighting regulations have since dimmed the glow.

atellite trails mark this photograph of the
vening sky *(below)* over the Anglo-Australian
bservatory 200 miles northwest of Sydney. Any
stronomical photograph exposed for ninety
ninutes or more is likely to show such streaks. More
ian 7,000 artificial objects orbiting the Earth are large
nough (over four inches) to be tracked by the North
merican Aerospace Defense Command *(bottom)*; only
ve percent are working satellites.

Reactors in Space

High-frequency radiation from satellites powered by nuclear reactors also threatens the new generation of orbiting astronomical observatories. Sensors aboard some of the observatories are designed to search for gamma ray bursts and other radiation from deep space. Unfortunately, nuclear reactors produce gamma rays during fission. They also fling antimatter particles called positrons into space; when the positrons come into contact with matter, they are annihilated, creating gamma rays in the process. Because of this, the Solar Maximum Satellite, in orbit since 1980, has steadily reported false gamma ray data. The 1990 Gamma Ray Observatory mission will be even more susceptible to nuclear pollution.

Dozens of radioactive satellites enmesh the Earth with their orbits *(below)*. At right, a hypothetical nuclear satellite emits positrons *(green dots)* along a magnetic field line *(purple band)*. The positrons in turn send gamma rays *(wavy lines)* toward a passing scientific satellite.

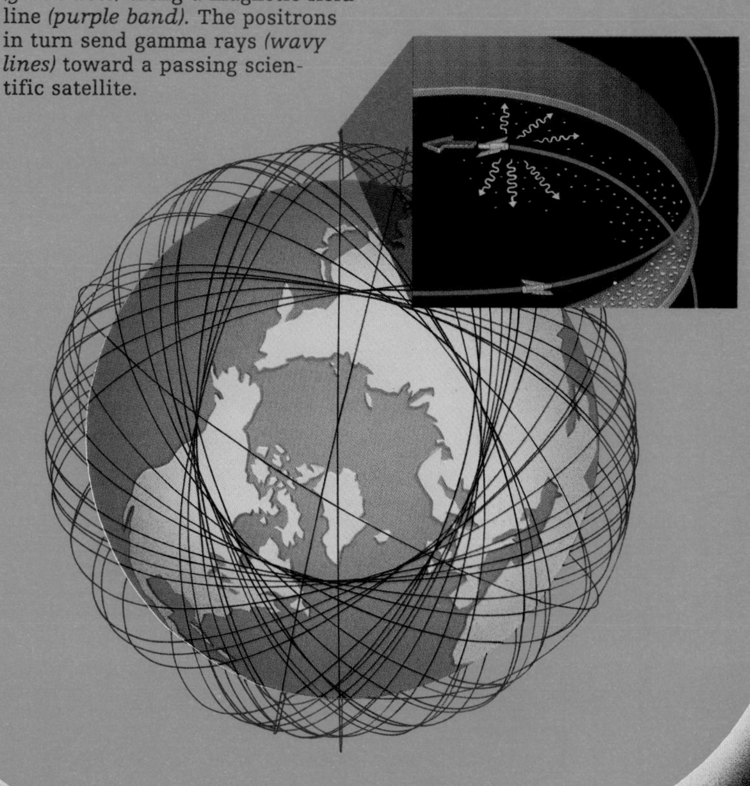

Drowning Out Cosmic Radio Signals

The revolution in electronics that has made life so much easier for the average consumer has seriously hindered radio astronomy. Astronomers working with radio telescopes stagger under a barrage of signals from microwave transmitters, electronic auto ignitions, mobile phones, and automatic garage door openers, among other things.

Cosmic radio signals are incredibly weak—one-trillionth of a watt or less—so they are easily drowned in a sea of more powerful local noise. They reach Earth as both broad-band emissions, falling across a wide range of wavelengths, or frequencies, and narrow-band, registering as specific lines on the electromagnetic spectrum. Unfortunately, interference comes in the same forms. Broad-band interference can overwhelm a radio telescope receiver or, worse, cause subtle distortions in observation data. Narrow-band interference can mimic astronomical spectral lines.

International agreements protect crucial radio astronomy frequencies, especially the signals emitted by hydrogen atoms at 1,420 megahertz and those of four key hydroxyl molecules at frequencies from 1,612 to 1,720 megahertz. These frequencies are important because the hydroxyl molecule occurs in star-forming regions of the galaxy. Trespassing on this protected territory are new navigation satellites. The Soviet Glonass satellites, for example, beam radio signals to Earth near the 1,612-megahertz frequency, hiding the hydroxyl signals from astronomers. When the Glonass satellite system is complete in the early 1990s, some radio astronomers believe it will eliminate a vital source of data on star formation.

Astronomers and engineers hope to ameliorate the situation through techniques such as time-sharing certain radio frequencies. Time-sharing allocates a portion of each second to satellite transmitters and the remainder of the second to radio telescope receivers. Scientists thus can identify and ignore signals that come from the satellites. Another protection is to institute quiet zones, such as the 13,000-square-mile Radio Quiet Zone around the radio telescopes at Green Bank, West Virginia. Increased regulation, more efficient frequency use, and improved electronic filters may also help ease the problem. Given the extent of electronic pollution, however, some scientists are convinced that believe the best place for future radio astronomy is the far side of the Moon.

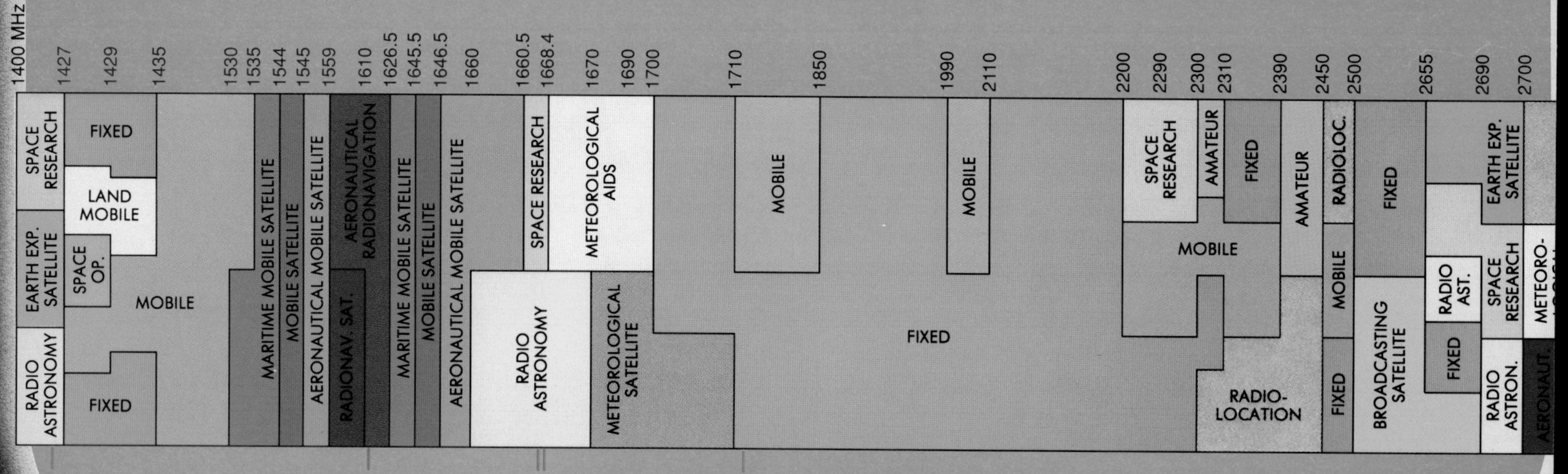

Hydrogen and hydroxyl signals (marked by red lines, above) must compete for air space with hundreds of other frequencies, shown here and on the following pages, that have been allocated to users in the United States. Although the wavelengths are regulated, signals that spill into adjacent channels are not. However, regulatory agencies try not to assign potentially troublesome frequencies to broadcasters that are located near radio telescopes.

method for using the Moon to find the exact position of 3C273. Training the 210-foot radio telescope at Parkes, Australia, on the radio source as the Moon passed in front of it, Hazard determined the exact moment when the radio signal was eclipsed. Since the position of the Moon at any time is precisely known, Hazard was able to locate 3C273 with great accuracy. He found that the source was split into two parts; the position of one agreed with that of a blue star, and the other seemed to be associated with a faint wisp of gas emanating from the star.

Hazard's observations prompted Dutch-born American astronomer Maarten Schmidt to look at 3C273 with the 200-inch telescope at Palomar Observatory in California. The object was so bright that he twice overexposed his photographic plates. Analyzing the spectrum of the light, Schmidt determined that the radiation was produced by a body composed of normal elements. But it was receding from the Earth at about 16 percent of the speed of light. Hubble's law, a fundamental principle of astronomy, indicates that recession velocities of such magnitude are due to the expansion of the universe, and that the velocity of an object is proportional to its distance from Earth. Schmidt's measurements placed 3C273 more than two billion light-years away, the most distant object discovered at the time. This was no star, nor even a normal galaxy, which at that distance would be only barely detectable with the technology then available. It was, by orders of magnitude, the most powerful source of radiation in the sky.

Schmidt's discovery set astronomy on its ear. A name was coined to describe the new phenomenon: quasi-stellar radio source, which evolved to quasar for short. Scientists the world over scrambled to detect similar objects—with great success. These remote powerhouses came to be understood as violently active galaxies that typically radiate the energy of hundreds or thousands of

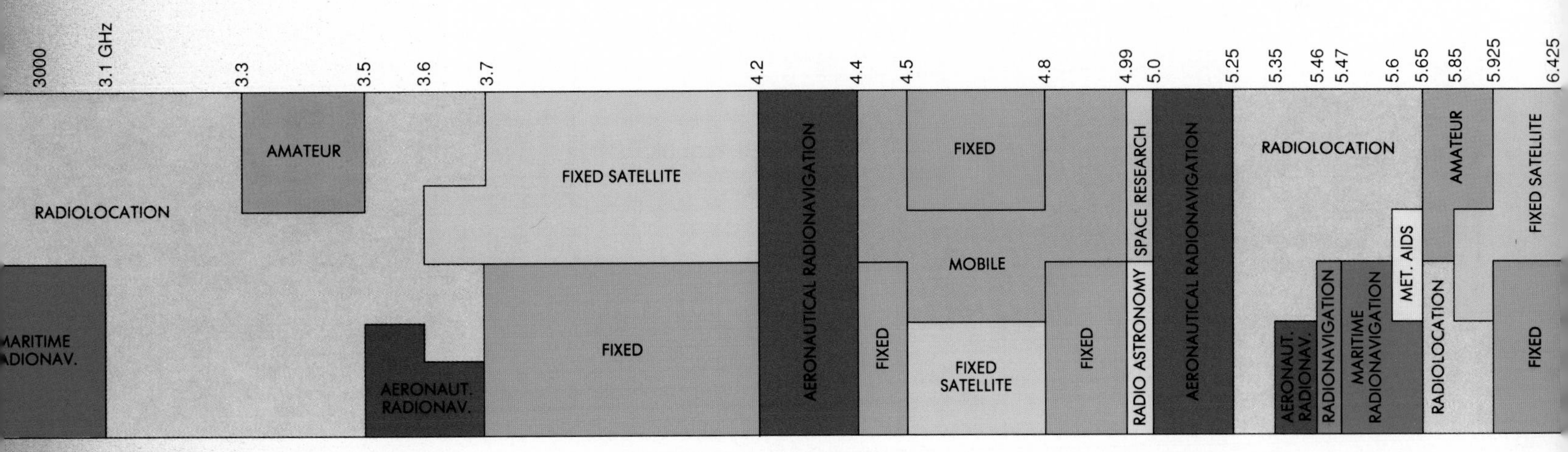

ordinary galaxies like the Milky Way, at wavelengths ranging across the entire electromagnetic spectrum. Most, however, gave the lie to their name, proving to be almost negligible transmitters of radio waves. Nonetheless, the collaborative effort behind the quasar discovery established a new trend in astronomy, serving as a model for other investigations that combined observations at different wavelengths to learn about various aspects of celestial objects. In subsequent years, radio and optical astronomers worked together not only to identify new quasars but also to learn more about such diverse phenomena as solar flares, the birth and death of stars, and interstellar clouds of molecules.

Eventually, many astronomers no longer identified themselves with a single type of radiation, as radio and optical astronomers had in the past. Instead, they tended to concentrate on a field of study, such as stars, galaxies, or interstellar gas. As new tools became commonplace, astronomers used whatever resources they needed to tackle the jobs at hand.

BIGGER, BUT NOT MUCH BETTER

In the early 1960s, the new tools still needed a great deal of improvement before they could measure up to the great optical telescopes. There was a fundamental technical problem to overcome in the case of radio telescopes. Radio waves are too long to focus clearly without huge antennas. For example, a 300-foot radio telescope operating at a typical radio astronomy wavelength of eleven centimeters gives a view of the sky that is more blurred than that perceived by the human eye at visible wavelengths.

Radio astronomers tried to overcome this unalterable fact of nature by building bigger instruments. In 1963, Cornell University astronomers assembled the world's largest radio telescope in a vast limestone hollow near

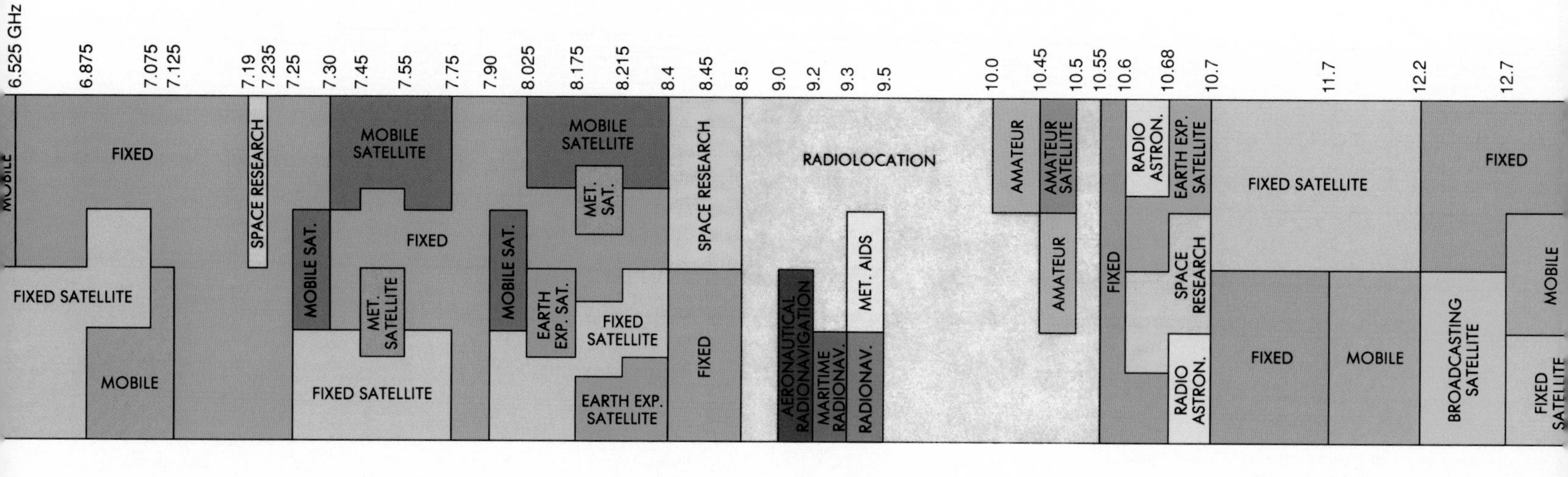

Arecibo, Puerto Rico. One thousand feet in diameter and more than 400 feet deep, the Arecibo telescope was immobile; it could scan the sky only as the Earth rotated and could look at only a narrow swath. Furthermore, even with its impressive area, Arecibo's resolution was no better than that of unaided human vision.

But greater size was not the only route to better resolution. Well before the Arecibo instrument was built, a team at Cambridge University began tinkering with an alternative technique: They applied the principles of interferometry to the creation of what became known as aperture synthesis telescopes. When two or more waves are added together, they produce a composite called an interference pattern. In an aperture synthesis telescope, more than one antenna is used to record radiation from the same source. The incoming waves are mixed in such a way that the interference pattern displays useful information about the strength and direction of the source.

The radio astronomers at Cambridge, headed by Martin Ryle (who later received the Nobel prize for his work), used special electronic circuits to combine the signals from several small, steerable telescopes. By mounting these dishes on a railroad track, observers could move them around, making readings at a variety of spacings. They could then merge all the observations, producing a signal comparable to one obtained from a hypothetical single telescope a mile or more in diameter.

The Cambridge aperture synthesis telescope—also known as a radio interferometer *(pages 34-37)*—became operational with two antennas in 1958; a third was added in 1964. It was a cumbersome device, with its sixty-foot dishes mounted on a mile-long east-west railroad track. Moving the antennas along the track was such a slow process that it took two months to complete a map of a single source. Nevertheless, the One Mile telescope, offering

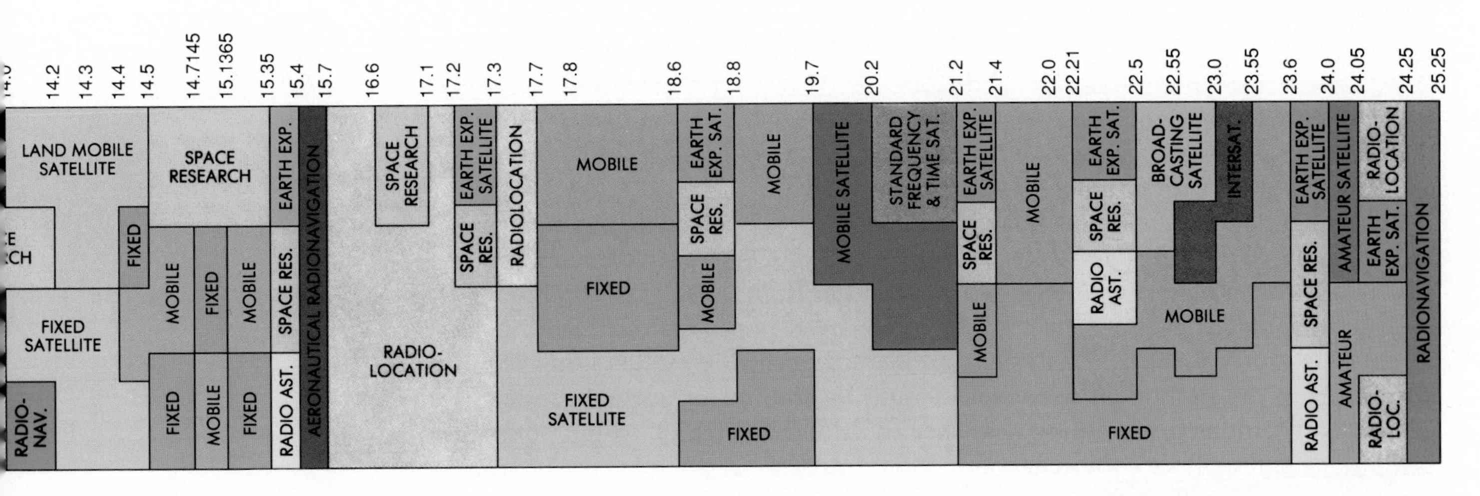

Astronomers today are focusing more attention on the highest frequencies. New technologies, such as satellite communications, threaten these top wavelengths.

resolution about thirty times better than any existing radio telescope's, sold the concept of interferometry to astronomers. Work soon began on larger, more flexible arrays, with the most ambitious plans unfolding at the National Radio Astronomy Observatory in the United States.

The NRAO was founded in 1957 near Green Bank, West Virginia, in a remote valley protected by the Allegheny Mountains from the electromagnetic noise of modern society. Operated by a consortium of universities and funded largely by the federal government, the observatory was able to build instruments far too expensive for the budget of a single institution. The first NRAO telescope was an eighty-five-foot dish, installed in 1959; it was augmented by a 300-foot dish in 1962. In that same year, David Heeschen, the first scientist employed by NRAO, became the facility's second director.

NEW STANDARDS

The soft-spoken Heeschen was already a veteran of radio astronomy. As a graduate student in the early 1950s, he had helped build the first radio telescopes at the Harvard Observatory in Massachusetts. One, a sixty-foot dish, was among the premier telescopes of its time and earned Harvard a reputation as a leader in the fledgling discipline. The NRAO, Heeschen decided, should also establish new standards for radio astronomy. In the early 1960s, he began working on plans to build an instrument with the ability to make radio images of objects of the same quality as those obtained with the largest optical telescopes. It would have twenty times the resolution of the Cambridge One Mile telescope—and would therefore require an array of movable telescopes spread across twenty miles.

The Very Large Array (VLA), as it came to be known early in the planning process, posed formidable technical and logistical challenges. The first hurdle was accumulating experience that could guide decision making about the size and configuration of the new facility. In 1964, Heeschen created a small interferometer at Green Bank by mounting a new telescope, identical to the existing eighty-five-foot dish, on a railroad track. Soon he added a third telescope on the track, as well as a smaller, portable antenna that could be set up more than twenty miles away.

Experiments with the Green Bank interferometer provided the basis for a VLA design proposal, prepared in 1967 by a team of NRAO astronomers and engineers. The antennas would be slightly smaller than those at Green Bank, but there would be twenty-seven of them, arranged in groups of nine along three arms of a Y. Each of the arms would be up to thirteen miles long. The signals from the individual telescopes would be channeled through underground waveguides to a building near the center of the Y, which would house computers and control equipment.

Finding a suitable location for such a huge array was not a trivial task. It would have to be at least twenty miles in diameter, relatively flat (to facilitate movement of the antennas), and far from large population centers. Searching topographical maps of the western United States, Heeschen's team noted

The world's largest radio telescope nestles in a natural hollow among the verdant tropical hills near Arecibo, Puerto Rico. A thousand feet in diameter, the dish is made up of 38,400 perforated aluminum panels. Cables strung from three towers around the rim support equipment that receives radio waves reflected up from the metal surface.

about forty potential sites. Subsequent inspection eliminated many of the forty. One proved to be covered with oil wells; another was an old lava bed, far too rough for vehicles; others were extremely isolated. Eventually a favorite emerged, an ancient lake bed in central New Mexico called the Plains of San Augustin. Easily accessible by a major highway, it was nonetheless removed from most human habitation: The two nearest population centers, ten miles in each direction, were old cow towns with a combined population of 1,300. Its altitude, 7,000 feet, was another plus, promising reduced atmospheric interference with celestial radio signals.

Five years of political wrangling and bureaucratic infighting followed the publication of the design proposal. The competition for scarce science funding was fierce, and more than once the VLA appeared headed for oblivion. The turning point came in 1971, when the National Academy of Sciences put out a document called *Astronomy and Astrophysics for the 1970s,* which labeled the VLA as the highest priority for astronomy in the United States. A year later, Congress approved $76 million to build the array.

The money was rationed at an annual rate of about $10 million, which forced an extended construction schedule. Although the delay gave Heeschen's team time to revamp their original design, by now years out of date, it also subjected the project to the ravages of inflation. A steep increase in the price of used railroad track in 1973 threatened to kill the VLA, which depended on what amounted to a small railroad, the nearly forty miles of track that would be used to move the antennas. Ever resourceful, the NRAO astronomers scrounged surplus rails from abandoned spurs on military bases around the country.

Scientific observations with the VLA began as soon as the first two antennas were installed in 1976; the first image was of the planetary nebula NGC 40. A year later, six antennas were available, and by 1980, the entire array of twenty-seven was in place. Operating twenty-four hours a day, it was an immediate success, detecting such phenomena as jets apparently shooting from the centers of radio galaxies, thin filaments of radio emission from the center of the Milky Way, and gravitational lens systems *(pages 66-69)* predicted years earlier by Albert Einstein. The VLA images had a resolution up to ten times that of the world's finest optical telescopes—performance that far exceeded its planners' hopes.

In the early years of operation, however, the facility's computer turned out to be a bottleneck, bogging down under the demand of transforming raw information—typically gathered at a rate of thousands of readings per second—into useful images. To cope with this load, NRAO astronomers and programmers developed the Astronomical Image Processing System (AIPS), a set of computer programs designed to handle every aspect of image processing. One program allowed observers to calibrate the signals received from their targets against the known strengths of several well-studied radio sources in the same area of the sky. Others helped weed out spurious data, whose unreasonably high or low values must have been caused by extraneous phenomena—atmospheric interference, for instance, or radio transmissions from artificial satellites. Yet another program filled in the gaps caused by the incompleteness of the array. The resulting images were incredibly crisp, showing sharp contrast and clear detail in areas that were hopelessly murky before processing.

Continuous improvements to the program, however, eventually exacerbated the very problem AIPS was intended to solve. By the late 1980s, AIPS had more than a half-million lines of computer code; the computing effort required for image processing ranged from ten to a hundred times what was needed to run the array and to produce the original data. This put such a heavy burden on the facility's computers (like everything else, limited by budget constraints) that some images were virtually impossible to resolve. Nevertheless, AIPS became an international standard, with many observers using it to process their VLA observations at other locations. Updated versions are made available on a quarterly basis to about 200 image processing centers worldwide for use on equipment ranging from small work stations to the fastest supercomputers.

BREAKTHROUGHS

Over the next decade, the VLA turned its versatile gaze on celestial phenomena of every size and distance. Solar researchers used it to study explosive outbursts on the surface of the Sun; planetary astronomers aimed it at Earth's neighbors, probing the atmospheres of Jupiter, Saturn, Uranus, and Neptune. Looking outside the Solar System, the VLA pinpointed powerful sources of radio waves in cool interstellar clouds where new stars are forming and mapped the expanding debris of exploded stars. By providing detailed portraits of other spiral galaxies, it gave astronomers new insights into the structure of the Milky Way. Some of the greatest breakthroughs with the VLA, however, have come from observing star systems very different from our own —radio galaxies and quasars.

Even the largest single radio telescopes show the quasar 3C273 as a pointlike source of radio waves. Optical telescopes detect an additional feature: a thin wisp of light apparently streaming from the central region of the quasar. The VLA image of 3C273 found an extended radio source with the same

Sunlight glints off more than two dozen eighty-two-foot radio telescopes at the Very Large Array near Socorro, New Mexico. Here, the VLA's dishes are pulled together into the array's most compact configuration, used for mapping large, faint sources. In its widest formation, with each arm lengthened to about twelve miles, the array is used to make detailed images of small sources.

Zeroing In on Radio Sources

Radio waves from celestial sources such as neutron stars and quasars arrive at Earth from all directions. To study a particular source in detail, radio astronomers must aim a telescope so that the waves emanating from the object arrive along the axis of the telescope's focus, perpendicular to the parabolic dish that serves as an antenna. In so doing, the waves that bounce up to the focal point all travel the same distance from the wave fronts *(solid lines, far right),* thereby arriving at the focus in phase, with their troughs and crests aligned. The closer the waves are to being in phase, the stronger the radio signal.

As with an optical telescope, the resolution of a radio instrument—that is, its ability to distinguish between closely spaced objects in the sky—improves as the diameter of the telescope's radiation collecting area increases in relation to the wavelength of the radiation being received. Generally speaking, as shown at right and below, the bigger the dish, the higher the resolution.

Whereas light waves are measured in millionths of an inch, radio astronomers deal with waves whose lengths range from fractions of an inch to yards. A 200-inch optical telescope can pick out an object the size of a dime at a distance of about two miles; to get an equivalent resolution from a radio telescope, the dish would have to be 150 miles in diameter. In view of the practical limitations on the size of a single dish, radio astronomers tackle the problem of improving resolution with a technique called interferometry—using many small instruments to synthesize the resolution of one impossibly large one *(pages 36-37).*

Blurry Vision

A radio telescope's images are based on the strength of the signals received at its focus, signals produced by radio waves arriving from the entire area of the telescope's field of view, or beam width. Beam width is inversely proportional to the size of the dish; the larger the dish, the smaller the beam width, and the better the ability to discern directional differences among incoming radio waves *(above, opposite).* If two sources are closer together than a telescope's beam width, the instrument cannot distinguish between them: It cannot narrow its beam to look at first one and then the other. Here, the beam width of a small radio antenna encompasses a region that contains two sources, emitting at thirteen and seventeen units of strength. Because the telescope can record only the total output from this region of sky—thirty units—the resulting radio image *(far right)* indicates just one source and gives no indication of its precise location within the region.

Going for Detail

A larger dish has better resolution because its greater span allows it to distinguish among incoming radio waves *(opposite, top).* In effect, it has a smaller beam width: It is capable of looking at a smaller portion of the sky at any one time. By scanning systematically across the same region of sky as that covered by the small dish above, the larger telescope measures the intensity of radio emissions in more discrete sections *(right).* Moving from left to right across the top of the region, it picks up no emissions in the first two squares, an intensity of thirteen units in the upper right corner, no emissions across the middle sections, and an intensity of seventeen units in the middle of the bottom row. When these numbers are converted into a radio image, the two radio sources appear separately *(far right).* However, the image does not reveal structure within the sources themselves.

When a dish antenna happens to be pointed directly at a celestial radio source so that the waves arrive perpendicularly *(dashed blue lines),* wave crests will be reflected in such a way that they arrive at the dish's focus at the same instant, reinforcing each other and producing a strong signal. Waves from a second source that do not arrive perpendicularly to the dish *(dashed red lines)* travel different distances from the wave front *(solid red line)* to the focus, throwing them out of phase and producing no signal. The bigger the radio dish is, the farther apart its edges are, and the better it can distinguish between signals that are perpendicular and those that are not—that is, the better its resolution.

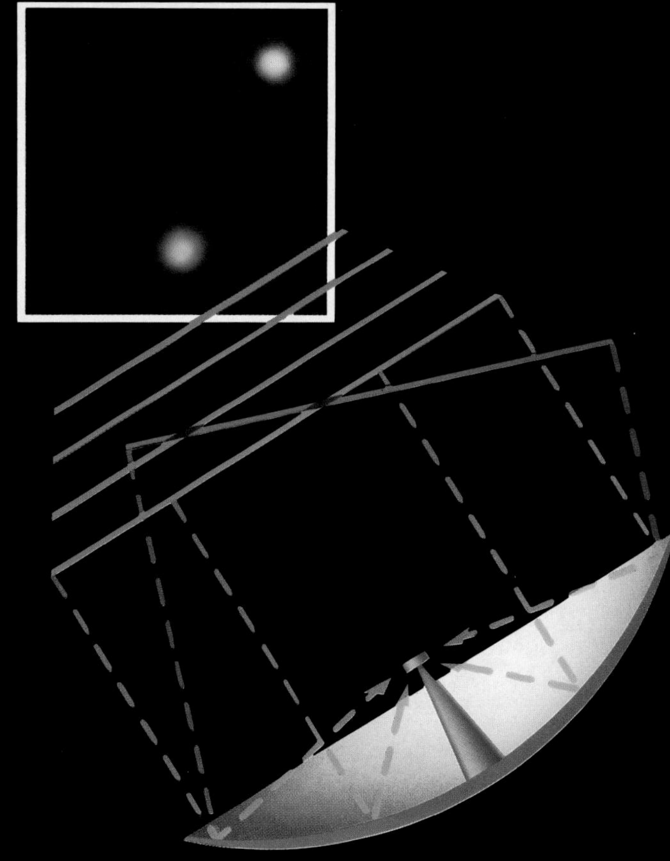

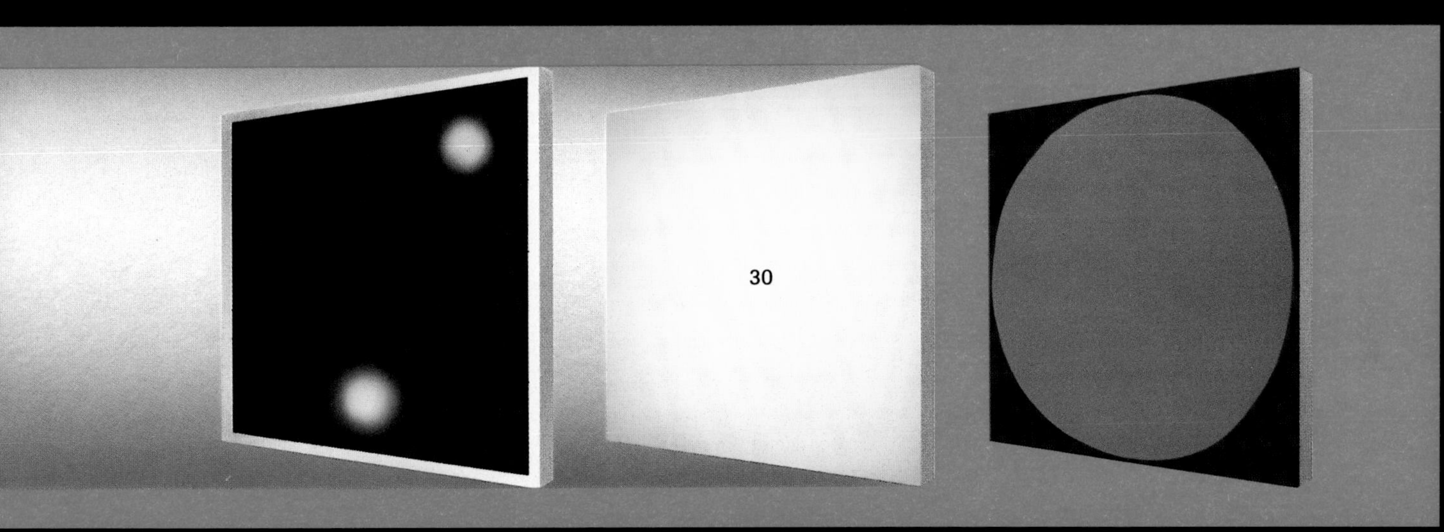

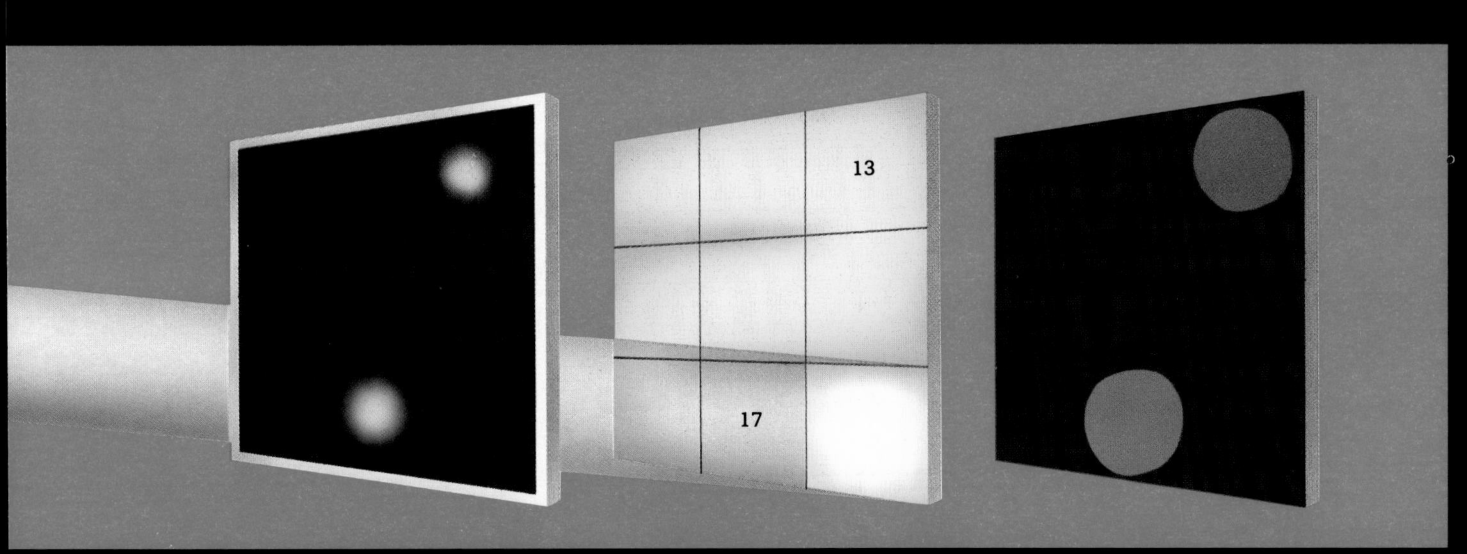

Synthesizing One Large Dish

Because radio waves are so much longer than those of visible light, even the world's largest single radio telescope—Arecibo, in Puerto Rico—has a resolution worse than that of the human eye. With the technique known as radio interferometry, however, astronomers can synthesize a dish that is hundreds—or even thousands—of miles across.

In effect, each dish in the array of telescopes that make up an interferometer acts like a small piece of the surface of the synthesized large dish. To create a focus for this piecemeal surface, astronomers link each telescope—using cables, microwave links, or optical fibers—to a central correlator, which combines the array of incoming radio waves into one signal. As with a single dish, a strong signal occurs when the waves travel the same distance to the focus, no matter which small telescope they passed through. Since the path lengths from the wave front through the telescopes to the focus are bound to be unequal, a computer is programmed to compensate by delaying certain signals so that all arrive in phase.

To fill the gaps between the telescopes, astronomers take advantage of Earth's rotation, which permits the radio interferometer to rotate in a circle beneath the area of sky that is under observation. With interferometers of fewer dishes, individual dishes can be moved on tracks to help fill in the gaps further.

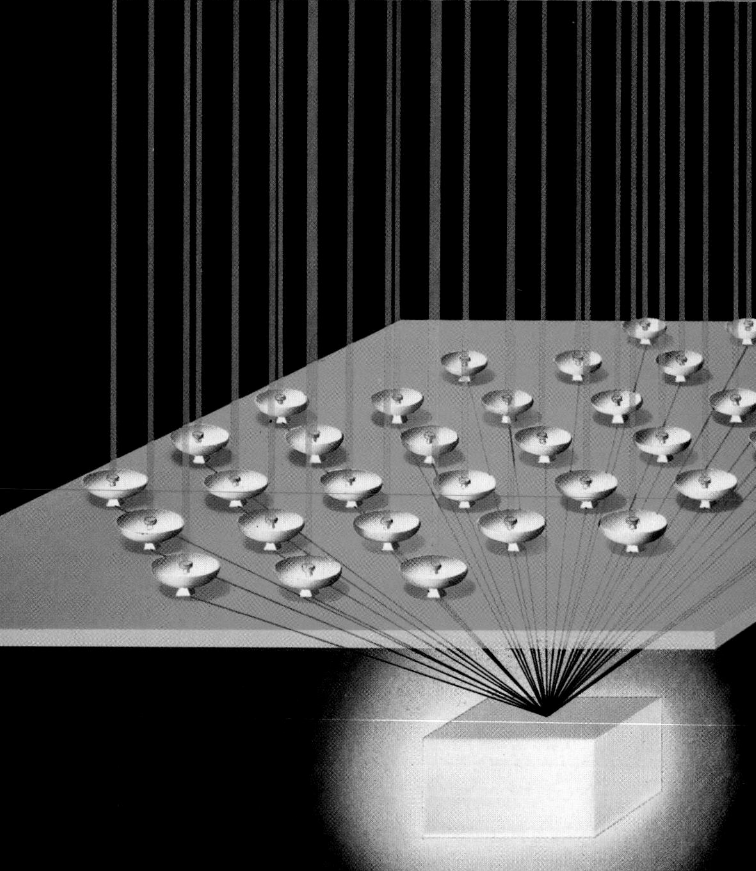

Sharpened Resolution

As the computer of an interferometer repeatedly shifts the array's field of view, or beam width (the larger the array, the smaller the beam width), the changing signal intensity is registered as numbers on a grid. So precise is the pointing procedure that the interferometer can not only zero in on the positions of two adjacent celestial objects but also show the distribution of radio energy around each of them. When the numbers are translated into a color-coded radio image *(far right)*, designated colors correspond to different intensities: orange for the strongest radiation, pale yellow-green for the weaker.

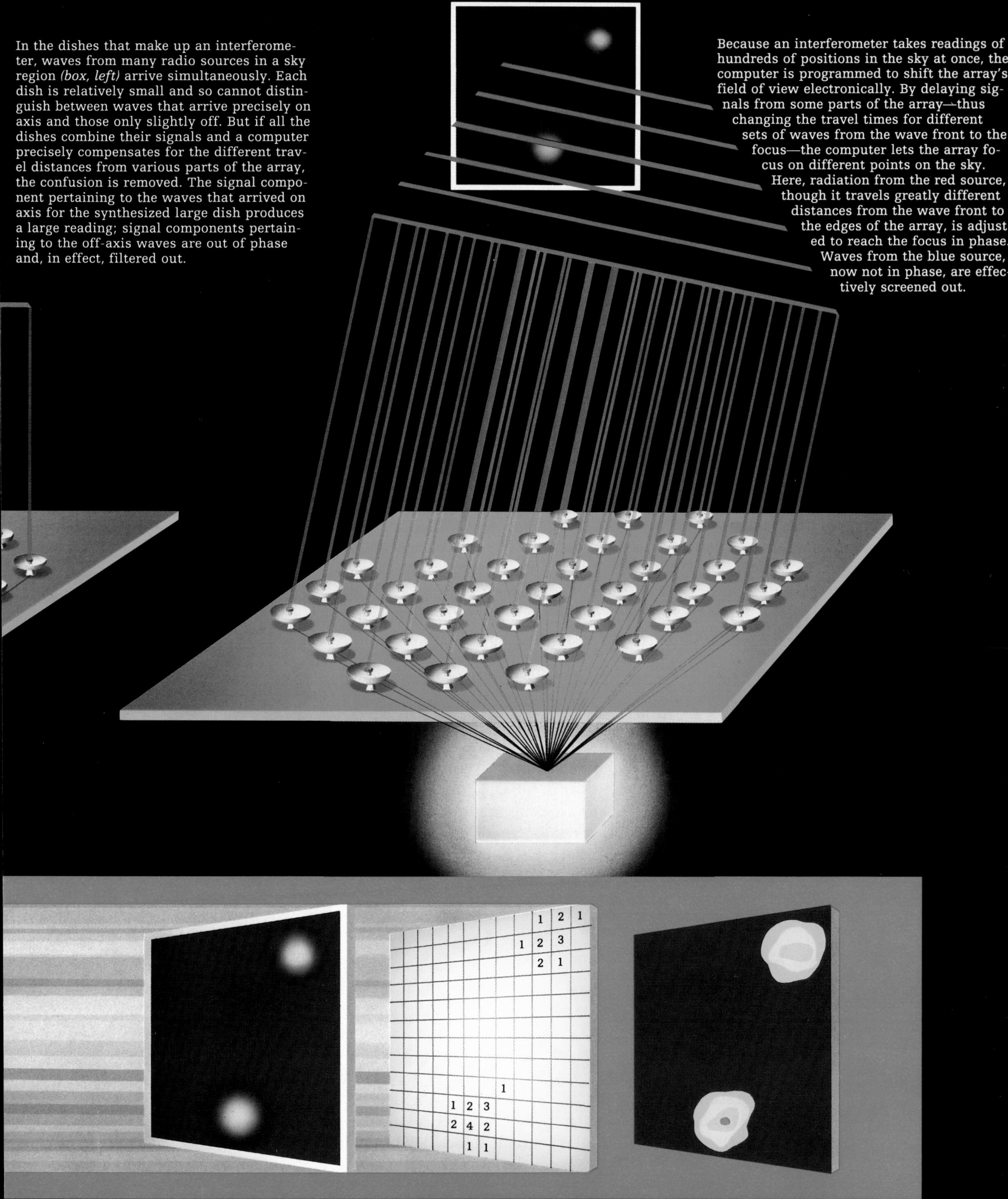

In the dishes that make up an interferometer, waves from many radio sources in a sky region *(box, left)* arrive simultaneously. Each dish is relatively small and so cannot distinguish between waves that arrive precisely on axis and those only slightly off. But if all the dishes combine their signals and a computer precisely compensates for the different travel distances from various parts of the array, the confusion is removed. The signal component pertaining to the waves that arrived on axis for the synthesized large dish produces a large reading; signal components pertaining to the off-axis waves are out of phase and, in effect, filtered out.

Because an interferometer takes readings of hundreds of positions in the sky at once, the computer is programmed to shift the array's field of view electronically. By delaying signals from some parts of the array—thus changing the travel times for different sets of waves from the wave front to the focus—the computer lets the array focus on different points on the sky. Here, radiation from the red source, though it travels greatly different distances from the wave front to the edges of the array, is adjusted to reach the focus in phase. Waves from the blue source, now not in phase, are effectively screened out.

orientation, which astronomers interpreted as a thin jet of high-energy electrons, moving at nearly the speed of light and emitting radiation as they spiral through a magnetic field. Similar jets have been observed in connection with so-called radio galaxies, sometimes extending hundreds of thousands of light-years beyond the galaxies themselves before billowing out into vast radio-emitting lobes.

Astronomers studying these mysterious structures believe they are caused by a supermassive black hole near the origin of the electron jets. So dense that not even light can escape its gravitational pull, a black hole inexorably draws in all matter within thousands of light-years. As stars and gas clouds swirl into the maw of the black hole before being swallowed, they are compressed into an accretion disk, where enormous thermal and magnetic forces blast a small fraction of the falling matter away from the center.

Unfortunately, a black hole is by its very nature invisible at all wavelengths; only indirect signs can confirm its existence. The detection of an accretion disk in 3C273, for example, would be persuasive evidence of the presence of a black hole. But from a distance of two billion light-years, even the VLA can make out details no smaller than 1,000 light-years across. An accretion disk at the center of 3C273 would likely be only a few light-years in diameter—undetectable until much finer resolution is available.

Using a variation on the theme of the VLA known as very long baseline interferometry (VLBI), astronomers have begun to peer deeper into the heart of 3C273 and others of its ilk. Six times a year, one or two weeks are set aside at six or more radio telescopes in the United States for VLBI observations. As each telescope points at the same cosmic source, the radio waves are recorded on magnetic tape together with timing signals from atomic clocks. The tapes are then transported to a central facility where they are synchronized and combined to form the image. In Europe, astronomers from several nations schedule similar cooperative efforts on a regular basis, and at times a number of European and North American antennas are combined to form an array as large as the Earth—with resolution equivalent to the ability to read this book from a distance of 3,000 miles. VLBI observations of 3C273 show that its radio jet is only a few dozen light-years long—far smaller than the optical jet, and striking confirmation that an exceedingly compact object is responsible for the prodigious power of the quasar.

YET ANOTHER GENERATOR

Even as evidence piled up of a black hole at the center of 3C273, radio astronomers were discovering another phenomenon capable of generating huge quantities of energy. It was called a starburst, the relatively sudden formation of many stars in close proximity within a giant cloud of gas and dust. Star formation is one of the enduring mysteries of astronomy; the road to even a partial understanding of starbursts was long and tortuous.

In 1963, American astronomer Sander Weinreb and several colleagues discovered hydroxyl radicals, the first molecules found in space. Hydroxyl (also

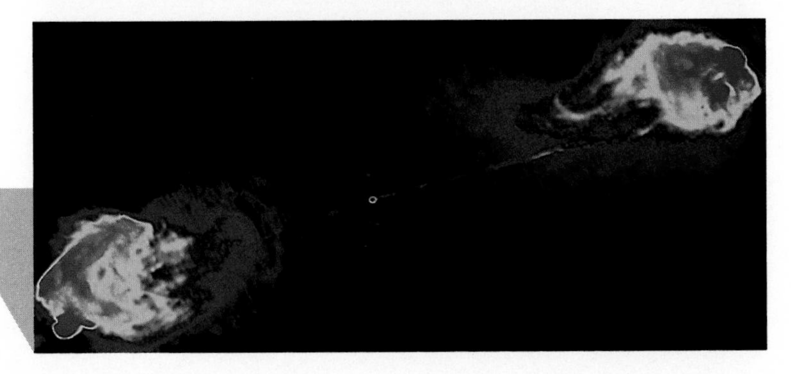

A VLA image of the radio galaxy Cygnus A is subjected to computer processing in the three-part sequence above. The top frame incorporates all the raw data—more than 2.5 million measurements—from about thirty hours of observation. The middle frame shows the image as it emerges from a computer program that creates the smoothest possible picture consistent with the original data. The computer then removes errors introduced by the antennas and electronic systems, producing the finished image at bottom.

known by its chemical name, OH), a water molecule with one hydrogen atom missing, radiates at a wavelength of eighteen centimeters, comfortably within the radio range. The discovery prompted a search for other molecules in space, but astronomers found that existing radio telescopes were inadequate to the task. Laboratory experiments showed that the most likely interstellar molecules, simple compounds of carbon and oxygen, would radiate at the shortest radio wavelengths, measured not in centimeters but in millimeters. These very short waves are easily absorbed by water vapor in the atmosphere. Furthermore, they can be focused only by telescopes whose surface imperfections are much smaller than the waves themselves.

For all their precision, ordinary radio telescopes are too rough to be effective collectors of millimeter waves. Therefore the NRAO decided in 1965 to add a specialized telescope to its inventory. Planners selected a site above most atmospheric water vapor, at an altitude of 6,300 feet near the top of Kitt Peak in Arizona. They designed a thirty-six-foot telescope with a parabolic collecting surface machined to tolerances of three-thousandths of an inch. Unlike most radio dishes, the telescope would be housed in a dome, for protection from distortion-inducing wind pressure and solar heating.

The millimeter-wave telescope was completed in 1967, but the first breakthroughs came in 1970, with the addition of a sensitive new radio receiver. Arno Penzias and Robert Wilson, two radio astronomers associated with Bell Telephone Laboratories, built the receiver by adapting technology developed by the phone company for digital telephone communications. Connecting it to the telescope, they settled in for what they expected would be the first of many long sessions of scanning the sky, looking for interesting objects. To their surprise, they almost immediately detected clouds of carbon monoxide in the Orion nebula.

In the intense period of observation that ensued, Penzias and Wilson used the powerful new receiver to discover a wealth of molecules where none had been seen before. They appeared to be concentrated, with the carbon monoxide, in thousands of huge gas clouds, up to 300 light-years in breadth, scattered throughout the spiral arms of the Milky Way. With enough matter to make several billion Suns, the clouds account for a substantial proportion of the galaxy's mass. Moreover, many of their constituents are organic chemicals, carbon-based molecules far more complex than astronomers had believed existed in space—compounds that could be the building blocks of life. More than sixty different types of molecules have since been discovered in molecular clouds, including water, ammonia, acetylene, formaldehyde, and ethyl alcohol.

The initial observations of carbon monoxide in Orion produced another surprise for Penzias and Wilson. Some of the gas was moving at about eighty miles per second, a much higher velocity than expected. Further investigation showed that these motions typically occurred around very young stars, which pushed out streams of gas in two distinct, opposed lobes. Bipolar flow, as it was called, was soon discovered in clouds throughout the Milky Way. Because the phenomenon was closely associated with the newest stars, astronomers studied it in hopes of gaining insights into star formation. Not even the millimeter-wave telescopes of the 1980s, however, could resolve enough detail to clarify the mechanics of the bipolar flows.

A new class of millimeter-wave interferometers, antenna arrays similar to the Very Large Array, may provide some answers. By the late 1980s, separate facilities were under construction at observatories in the United States, Spain, and Japan. But more information about nascent stars was already available at other wavelengths, particularly infrared.

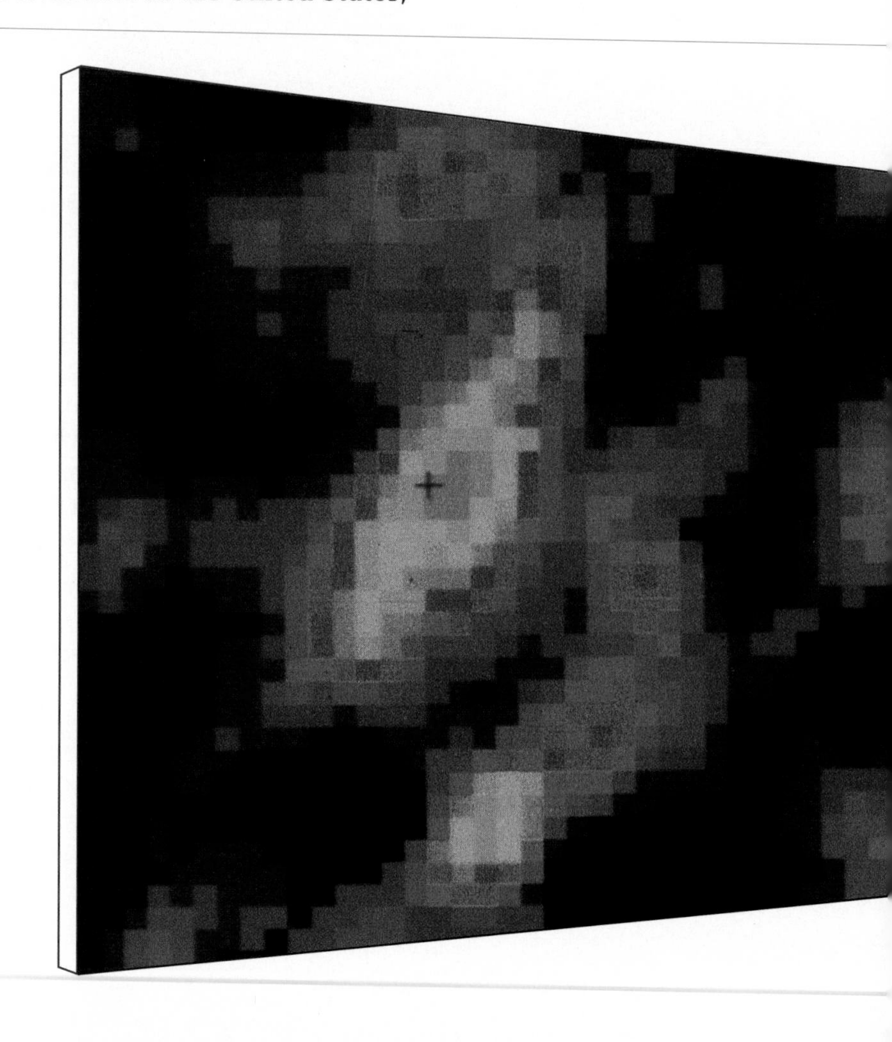

HAMPERED PROGRESS

Infrared astronomy made slow progress through the first half of the twentieth century. In the 1920s and 1930s, near infrared radiation was detected from the planets and some bright stars; by the 1950s, efforts were under way to detect organic molecules on the surface of Mars, through their infrared emissions. But attempts to push observations to longer wavelengths continued to be hampered by inefficient instruments that were simply not sensitive enough to detect the feeble emanations of distant bodies as cold as twenty degrees Kelvin.

The solution to the problem came from a corporate research lab with no connection to astronomy. Frank Low, a physicist at Texas Instruments in the early 1960s, was working on applications of a phenomenon known as superconductivity when his supervisor suggested that it might be useful in building a sensitive bolometer. Many materials, when cooled to very low temperatures, become so-called superconductors—that is, they conduct electricity without resistance or loss of energy. One of the greatest practical difficulties associated with superconductivity, however, is that a leak of energy into the cooled system can heat the conductor above the temperature that marks the transition point back to ordinary conductivity. The idea that Low pursued was to cool a thin strip of material such as aluminum just below its transition temperature and position it in a can, behind a small opening. If a small amount of energy came through the hole and struck the aluminum, it should warm the metal above the transition temperature. The resulting change in the conductivity of the strip would signal the arrival of the radiation.

As it happened, the superconducting bolometer did not work as planned. But Low decided to try another tack. Texas Instruments was a leader in research on semiconductors, materials whose conductivity can be altered by a variety of factors. Low set to work on a semiconductor bolometer using germanium, a substance that can be prepared so its electrical conductivity changes when it absorbs energy. The general principle was the same as for the superconducting device: Changes in the conductivity of the germanium would be directly proportional to the amount of energy entering the bolometer. Low completed the instrument in just a few weeks and found that it worked better than he had hoped, proving to be at least twenty times more sensitive than any existing bolometer.

In succeeding months, Low carried out more experiments, developed the theory behind the device, and wrote a paper describing it for the *Journal of the Optical Society of America.* After publication of the article in late 1961, some astronomers visited Low's lab to learn more about the new instrument, which they encouraged him to adapt for astronomical use. Low took their blandishments to heart and soon came up with another important invention.

The germanium bolometer worked best at a temperature near absolute zero, maintainable only by surrounding it with liquid helium, at four degrees Kelvin. Extremely cold liquid was commonly held in insulated flasks called Dewars (after their inventor, Scottish physicist James Dewar). These double-walled vacuum bottles, similar to the familiar thermos bottles used to keep coffee hot, were too fragile for use in observatories, not to mention aboard the balloons and rockets that were necessary for getting above Earth's infrared-sapping atmosphere. Furthermore, existing Dewars tended to leak, requiring regular refills of expensive liquid helium. Low developed a strong, efficient metal Dewar that made his bolometer a practical astronomical instrument.

Radio astronomer Frank Drake was particularly interested in Low's inventions, which Drake believed could be used to detect millimeter waves. Following a visit to the Texas Instruments laboratory, Drake returned to his post at the NRAO, where he persuaded David Heeschen to offer Low a job and a substantial budget for further development of the bolometer. After some soul-searching, the physicist decided to give up his promising career in the

disk of carbon monoxide gas round the young star HL Tau hows up as concentric regions of ed, orange, yellow, and green in his image from the Owens Valley nillimeter-wave interferometer n California. The carbon monoxde cloud, which also contains arge quantities of dust, may be precursor to a planetary sysem around the star.

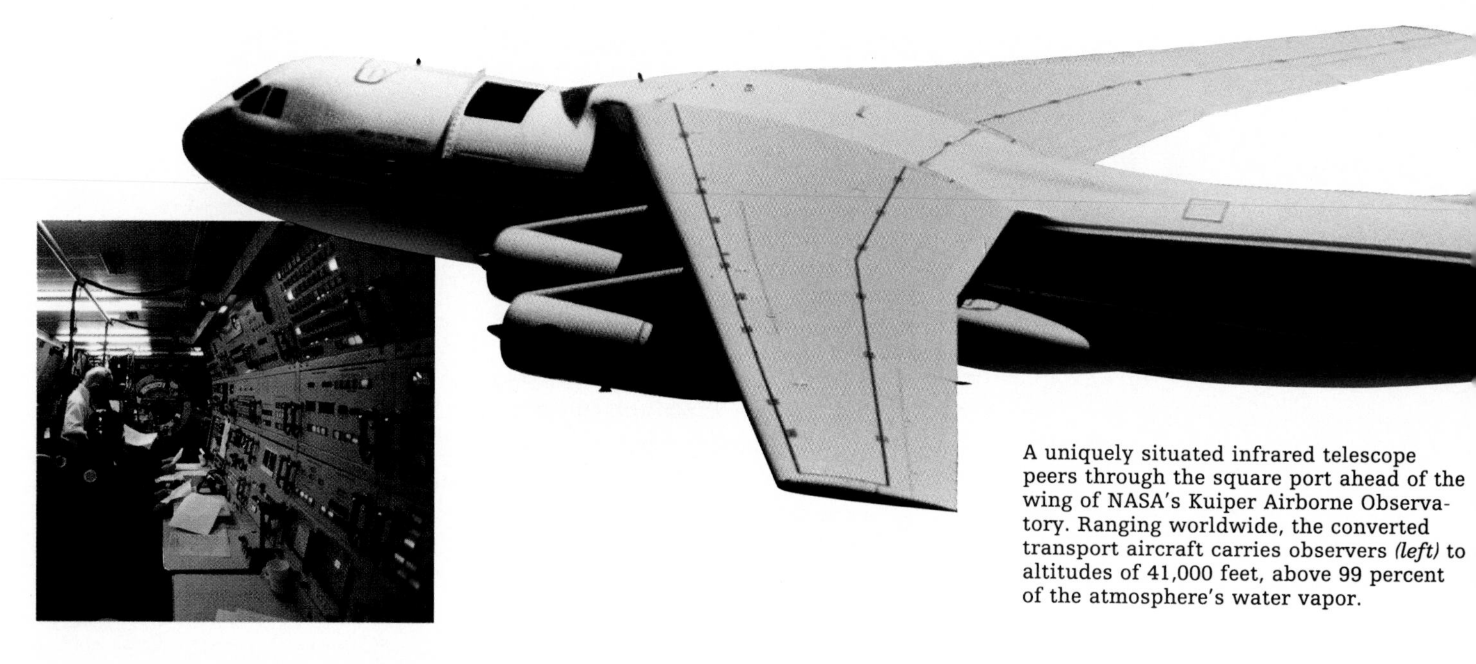

A uniquely situated infrared telescope peers through the square port ahead of the wing of NASA's Kuiper Airborne Observatory. Ranging worldwide, the converted transport aircraft carries observers *(left)* to altitudes of 41,000 feet, above 99 percent of the atmosphere's water vapor.

electronics industry for an uncertain future in astronomy, a field in which he had no experience.

The experience was not long in coming. Low moved to the isolated NRAO facility at Green Bank, where he found an excellent library and some of the world's leading radio astronomers. With little to do in the evening but read in the library or talk with coworkers, Low quickly came up to speed in his new trade. In 1963, one of his bolometers rode a balloon to 78,000 feet to make infrared observations of Mars—and deflated the notion, held by some astronomers, that the radiation emanated from living organisms on the Red Planet.

Low also worked on the Kitt Peak millimeter-wave telescope, and moved to Arizona to help with the installation. By 1966, he had decided to stay out west rather than return to Green Bank when the work was completed. For the next few years, Low divided his time between positions at the University of Arizona in Tucson and Rice University in Houston, Texas. With his now well-developed instruments, he became a pioneer in infrared observation. Atop rockets, beneath balloons, and attached to ordinary optical telescopes, the bolometers traced the dusty detritus of cosmic phenomena—in broad rings around stars, clouds near the nuclei of galaxies, and halos around quasars.

One of Low's most important contributions was the 1967 discovery of a giant cloud of gas and dust in the constellation of Orion. Working with Douglas Kleinmann, a graduate student at Rice, Low used his own bolometer and the Kitt Peak telescope to find that the cloud, with a mass just 200 times that of the Sun, outshines Earth's local star by a factor of 100,000. Unlike the Sun, however, the Kleinmann-Low nebula, as it was named, emits almost all of its radiation in the infrared bands; some radio waves are detectable, but at optical wavelengths the source is invisible.

With such an extraordinary outpouring of infrared radiation begging for explanation, a number of astronomers turned their attention to the nebula.

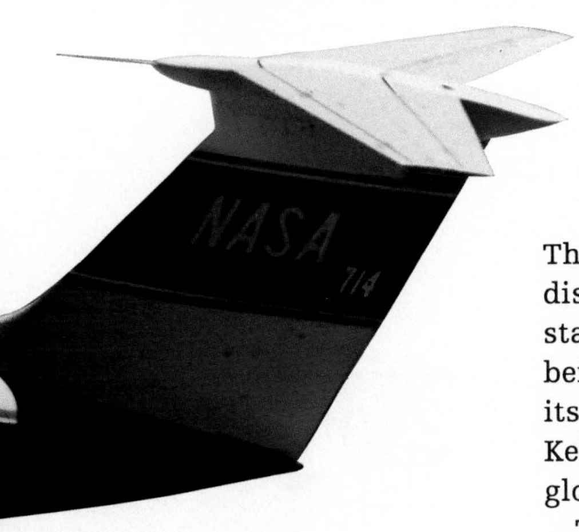

They soon determined that it was a region of active star formation at a distance of 1,600 light-years, one of the nearest ever found. A bright young star, dubbed IRc2, appeared to be the source of a large fraction of the energy, being half as bright as the whole nebula in the near infrared. Furthermore, its spectrum indicated that IRc2 is radiant enough to heat dust to 150 degrees Kelvin across the one-light-year breadth of the cloud, causing the infrared glow visible from Earth.

The brilliant star also showed bipolar flow at millimeter wavelengths, leading to speculation that it was compressing the surrounding gas and triggering the formation of other stars. One object within the nebula appeared to support this idea, emitting infrared radiation at the 600-degree temperature characteristic of a protostar with no energy-producing core. As it turned out, later spectral analysis indicated that the object was a little farther up the evolutionary ladder: It was a very young star, no more than a few thousand years old. The infrared radiation came from a ring of warm dust about the size of the Solar System—a remnant of the star's birth.

DETECTORS—AND CONTROVERSY—ON THE RISE

By the early 1970s, infrared detectors were being pointed in every direction, and provoking controversy nearly everywhere they turned. Of particular interest was the bright infrared radiation emanating from the centers of most galaxies, including the Milky Way. Concentrations of old stars might produce a small part of such radiation, but astronomers differed sharply as to the origin of the rest. Some said it must come from a relatively recent starburst, while others argued that the energy was produced by matter heating up as it fell into an accretion disk around a supermassive black hole. The discovery in the 1970s that quasars and other violently active galaxies are also strong sources of infrared radiation only heightened the debate.

These arguments about the physics of remote objects were based on incomplete information. Most of the available data came from observations in the near infrared, which were relatively easy to make from the ground and therefore quite numerous. Longer-wavelength infrared was much harder to capture, with such expedients as airplanes, rockets, and balloons offering only short, expensive looks at the sky. Many scientists began to think of installing an infrared telescope entirely above the absorbing blanket of the atmosphere, on a satellite in Earth orbit.

In the mid-1970s, Low joined forces with several other distinguished infrared astronomers to propose just such an orbiting observatory to the National Aeronautics and Space Administration, which had already sent up a number of astronomical spacecraft. From NASA's standpoint, the timing was abysmal: The agency was in the midst of a budgetary crisis, and other projects had much higher priority. The proposal probably would have gone nowhere except for the intervention of the Netherlands Agency for Aerospace Programs. The Dutch, as it happened, already had a design for an orbiting telescope. Because the Infrared Astronomical Satellite *(IRAS),* as they called

it, was proving more expensive than they had originally planned, the Dutch asked NASA to collaborate.

NASA accepted the Dutch offer in 1975, and the partnership was joined in 1977 by Great Britain. The British were to build the ground station and control center for the satellite, and the Dutch were responsible for the spacecraft itself, including electrical systems, on-board computers, and equipment for aiming the telescope. The American team would construct the telescope, the infrared detectors, and the cooling system, and would conduct the launch.

The design called for a spacecraft that would be about as bulky and heavy as a medium-size automobile. The telescope would have a twenty-three-inch mirror and an array of detectors sensitive to most of the infrared portion of the electromagnetic spectrum, and it would be housed in one of Low's Dewars, filled with 127 gallons of liquid helium. All this would ride into a 560-mile-high polar orbit atop an American Delta rocket, with the launch date set for August 1981.

Building a high-quality telescope is never easy, even for installation on solid ground. The task requires not only technical skills but also the political savvy to smooth out bureaucratic bumps and keep large work teams functioning well. Anyone associated with a space observatory should also have the nerves of a riverboat gambler, since even the best-built telescope can evaporate aboard an exploding launch vehicle or succumb to mysterious malfunctions in space.

The nerves of the *IRAS* team were severely tested as its members struggled to shape their dream into a reality. One day before the end of the mirror-polishing process, a tool broke and gouged the smooth disk; additional polishing repaired some of the damage but brought the mirror only up to the minimum performance specifications. When the original semiconductor detector array proved to be unreliable, the launch was postponed so that a "tiger team" could undertake a crash program to build new detectors designed by Frank Low.

Complications kept cropping up even after the telescope was mounted on the spacecraft and filled with liquid helium. Some yielded to simple electronics fixes, but others could have been dealt with only by postponing the launch yet again to dismantle the satellite for repairs. By now the spacecraft was getting old, and scientists feared that other vital parts might deteriorate in the span of several months required to deal with the present problems. Deciding that the risks of waiting were greater than the chance that one of the last-minute glitches would cripple the satellite, NASA officials set January 25, 1983, as the date when *IRAS* would be launched from Vandenberg Air Force Base in California.

The team's luck changed with the launch. *IRAS* took off without delay

Regions of active star formation blaze white in an infrared picture of the Large Magellanic Cloud, a small galaxy gravitationally bound to the Milky Way. Reconstructed from hundreds of narrow strips recorded by the Infrared Astronomical Satellite, the image shows dust particles heated by older stars as regions of yellow; blue circles are foreground stars within the Milky Way.

A large yellow splotch marks dense concentrations of stars, gas, and dust around the galactic center in this infrared view of the plane of the Milky Way. Giant dust clouds, heated by emissions from nearby stars, show up as yellowish knots that are scattered along the band.

during a six-hour stretch of good weather between two of the severe storms that ripped the California coast that winter. For the next ten months, the satellite performed almost flawlessly, scanning more than 95 percent of the sky four times and detecting about 250,000 celestial sources of infrared radiation before its supply of cooling helium was exhausted. Some of the sources were familiar—the center of the galaxy, the Kleinmann-Low nebula, stars, galaxies, and quasars such as 3C273. Others were new, including six comets and an asteroid that may pass between Earth and the Moon in the year 2115.

One of the most exciting discoveries was an old object, now seen in a new light. Vega, a bright star about twenty-six light-years from Earth, is familiar to amateur astronomers and navigators as well as to professionals, who use it for testing telescope performance. Early in the *IRAS* mission, ground controllers directed the satellite to scan across the position of Vega to check whether the telescope and detectors were working as expected. They were startled to find that the infrared radiation at longer wavelengths was much stronger than prevailing theories said it should be.

Vega's far infrared radiation apparently originates in a disk of dust grains a millimeter or more in diameter, heated by the star to about eighty-five degrees Kelvin. Scientists believe that these grains may be left over from the cloud of dust and gas that formed Vega a few million years ago. A similar but much smaller disk is thought to have existed around the Sun after its birth, providing the raw material from which Earth and the other planets were formed. The disk around Vega could be the progenitor of future planets; some asteroid-size bodies may be forming there even now. Detecting planets at Vega's distance, however, is still beyond the capabilities of even the best infrared telescopes.

Another object *IRAS* examined closely is a strangely shaped galaxy named Arp 220 (from its listing in astronomer Halton Arp's *Atlas of Peculiar Galaxies).* Optical images of Arp 220 show loops of luminous material apparently issuing from two bright oval-shaped smudges. *IRAS* revealed that the galaxy was emitting about eighty times more energy in the far infrared than at all other wavelengths, making it one of the most luminous galaxies ever observed, comparable in total energy output to quasars.

Many astronomers believe that Arp 220 is actually two galaxies in collision. In such a spectacular event, unfolding over millions of years, the widely spaced stars of the individual galaxies are unlikely to smash into each other. That fate is reserved for clouds of dust and gas, which are then racked by powerful shock waves. Heated by compression, the gas radiates furiously; the dust absorbs most of the radiation and reemits it at infrared wavelengths. In portions of the clouds where the density is pushed high enough, millions of massive stars may form in the brief period of a few million years.

Gas clouds in such an interaction may have another indirect, but ultimately dramatic, effect on the galaxies. Losing their orbital momentum, the clouds would begin to spiral toward their galactic nuclei. The resulting concentrations of gas could further enhance the central starburst or become fodder for a supermassive black hole in the nucleus. According to one astronomical model of Arp 220, gas is heated by both a starburst and an accretion disk around a central black hole. Eventually the gas that does not fall into the accretion disk around the black hole will be blasted away by supernova explosions, stellar winds, and pressure from the intense radiation fields. At this point, according to the model, as the infrared radiation diminishes and the activity around the black hole becomes dominant, a quasar will become visible through the clouds.

Regardless of the accuracy of this particular model, most astronomers are convinced that ultraluminous infrared galaxies like Arp 220 hold keys to understanding the energy source for quasars. A new generation of telescopes, on the ground and in orbit, may help to unlock these secrets. Scientists who are still gleaning important discoveries from *IRAS* data look forward to its successor, a satellite with a twenty-four-inch telescope to be launched by the European Space Agency in early 1993. The instrument will be sensitive to radiation across the full range of infrared, from near to far, and will carry enough coolant for eighteen months of operation.

BEYOND VIOLET

If satellites are important to infrared astronomers, they are essential to scientists observing at ultraviolet wavelengths. The ozone layer, ten to thirty miles above Earth's surface, screens out most ultraviolet rays, protecting terrestrial life from their invidious effects, which include sunburn and skin cancer in humans. This atmospheric shield makes Earth-based observations of ultraviolet sources almost impossible, forcing astronomers to loft their instruments to high altitudes.

As its name suggests, ultraviolet radiation occupies the portion of the electromagnetic spectrum just beyond violet light, the shortest-wavelength visible radiation. Material with a temperature between 10,000 and 1,000,000 degrees Kelvin emits most of its energy in the ultraviolet. The atmospheres of practically all stars fall in this range, as do the surfaces of massive stars, white dwarfs (small, dying stars), and hot regions of the interstellar gas. Furthermore, the most abundant atoms in the universe, including hydrogen, helium, carbon, nitrogen, oxygen, and silicon, have prominent spectral features in this wave band. Thus ultraviolet astronomy can provide invaluable clues, obtainable in no other way, about the temperature and composition of remote objects.

The potential value of an orbiting ultraviolet observatory was first proposed in 1946 by American astrophysicist Lyman Spitzer, Jr., in a report for

the U.S. Air Force on the scientific uses of satellites. When he joined Princeton University the next year as chairman of the astronomy department, Spitzer assumed a leading role in the campaign to put telescopes into space for the purpose of optical and ultraviolet observations. At first Spitzer and his colleagues had to depend on sounding rockets, small missiles that could carry lightweight payloads to the edge of space before falling back to Earth. For a decade, ultraviolet astronomers focused on the Sun; not until 1957 did they detect stellar ultraviolet radiation.

With the establishment of NASA in 1958, Spitzer's concept began to flourish. The space agency drew up a program of four Orbiting Astronomical Observatories, identical spacecraft that would carry different instrument packages to measure ultraviolet radiation from stars or interstellar gas. Spitzer led a Princeton group assigned to design a thirty-six-inch telescope for the fourth mission.

A foil-covered tube shields the ultraviolet telescope aboard the International Ultraviolet Explorer; winglike solar panels provide electricity needed to operate the spacecraft and its scientific payload. Launched in 1978 with a life expectancy of three years, *IUE* has survived several equipment failures—including the loss of four of its original six guidance gyroscopes—to become the longest-lived astronomical satellite.

The first satellite in the *OAO* series developed an electronic malfunction and stopped operating after only two days in orbit in 1966. Subsequent launches came at intervals of about two years. *OAO-2* used relatively small telescopes to carry out a general survey of the ultraviolet sky; its successor failed to reach orbit in 1970. Spitzer's spacecraft—christened *Copernicus* to honor the great Polish astronomer's 500th birthday—was the first to take close looks at specific ultraviolet targets.

Launched in 1972, *Copernicus* relayed data to Earth for nine years. Perhaps its greatest accomplishment was augmenting astronomers' understanding of interstellar gas in the Milky Way, still sketchy when the satellite went up. *Copernicus* showed that the gas was distributed in an irregular patchwork of dense molecular clouds, surrounded by a vast network of tunnels of hot, low-density gas blasted out by exploding stars. Later investigations showed that in some places, this hot gas seemed to billow far above the plane of the galaxy like a huge fountain.

Two years before it ceased operation, *Copernicus* was joined in orbit by the International Ultraviolet Explorer *(IUE)*, a satellite built as a joint venture by NASA, the European Space Agency, and the British Science Research Council. Equipped with more-efficient electronics, *IUE* was able to examine much fainter objects than any of its predecessors had, over a broader band of wavelengths. In 1989, *IUE* marked its eleventh year of operation, making it the longest-lived satellite observatory, and one of the most productive: Well over a thousand published scientific papers have described its discoveries.

Lofted into an orbit that keeps it stationed at an average altitude of 22,500 miles above the Atlantic Ocean, the satellite is in continuous contact with one or both of its ground stations in the United States and Europe. Astronomers use it much as they might a ground-based observatory, visiting a control center (the European facility is near Madrid and the other is at NASA's Goddard Space Flight Center outside Washington, D.C.), viewing the sky in the vicinity of their targets through optical cameras aboard the satellite, and then issuing the commands necessary to point the telescope in the desired direction.

Like almost every powerful astronomical instrument, *IUE* has been used to study the most energetic galaxies and quasars. It proved particularly useful for observing starburst regions in galaxies such as NGC 1068. At longer wavelengths, the case for the presence of hot, young stars is entirely inferential, obtained by measuring the radiation reemitted by the ultraviolet-absorbing shrouds of dust. *IUE,* however, detected the tiny fraction of the original ultraviolet radiation—less than one percent—that does escape the dust clouds in NGC 1068. The spectral signatures in this radiation unmistakably identified massive stars that were born less than a million years ago, yesterday in astronomical terms. Furthermore, *IUE* observations support the idea that the galaxy may be evolving into a quasar. As the starburst consumes most of its surrounding dust and gas, and blasts away the rest with its radiation, the brightness of the galaxy will increase manyfold.

Penetrating near the heart of a potential quasar, the technology of ultraviolet astronomy has confirmed and extended the findings at longer wavelengths, bringing astronomers one step nearer to understanding one of the most awesome objects in the cosmos. But the core of a quasar, the locus of its violent engine—probably a supermassive black hole—remains hidden. Like many other celestial phenomena, it encodes its innermost secrets in radiation whose wavelengths are shorter than ultraviolet. Detecting and interpreting such waves requires technical resources even more ingenious and precise than those that have opened so much of the universe in and around the range of visible light.

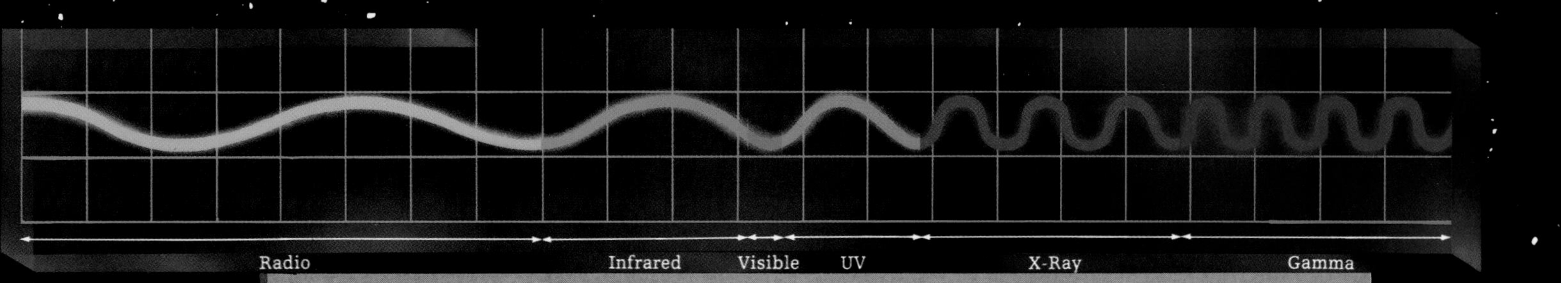

SIGNATURES OF THE STARS

The light from stars twinkling in the night sky is only a narrow sample of the electromagnetic radiation generated in the universe. Celestial objects—stars included—also emit nonvisible waves of electromagnetic energy when charged atomic particles collide, interact with magnetic fields, or encounter other radiation. Like light, these waves travel through space at about 186,000 miles per second.

Radiation has a peculiar dualistic quality that allows it to be detected not only as waves but also as particles, called photons. Astronomers who analyze radiation often combine the properties of both aspects, speaking, for example, of photons with a given wavelength. In any event, they use the measurable qualities of radiation—wavelength, frequency, and intensity—to determine the temperature, density, and composition of the matter that gives off radiation and to understand the events that trigger its release.

The spectrum of electromagnetic radiation, illustrated at the top of the page, extends from radio waves *(left),* as long as a few miles, to gamma rays *(right),* shorter than the radius of an atom. The visible, or optical, portion of the spectrum occupies a slender band of wavelengths near the center.

Virtually everything astronomers have learned about remote phenomena is a consequence of their ability to decipher the code of cosmic electromagnetic radiation. The following pages introduce some important analytical tools and present four examples of their use in unraveling astronomical mysteries.

CLUES IN THE RISE AND FALL OF PHOTONS

Every celestial body and event declares its nature in the wavelength and intensity of the electromagnetic radiation it emits. The photons of this radiation, with neither mass nor electric charge, remain largely unaltered as they travel through the universe. This makes them excellent sources of information: Because they reach Earth intact, their numbers and qualities can tell the tale of their origins.

Like detectives working on a complex case, astronomers rely on painstaking observation and laboratory work to identify and interpret the evidence. They begin an investigation by plotting a graph of radiation from a celestial body, showing the intensity of emissions at every wavelength. The name of this graph—spectral intensity curve—is commonly (and somewhat misleadingly) abbreviated to spectrum.

Most objects emit radiation over a broad range of wavelengths, with the intensity rising and falling across the spectrum. By comparing the resulting peaks and valleys with known patterns that correspond to physical processes, a skilled astronomer can piece together a picture of the radiation source. Three of the most common spectral shapes are shown at right, with explanations of the phenomena that cause them. Although the curves are illustrated separately here for clarity, a typical spectrum is more complicated, combining signs of several types of radiation.

One regular spectral component that does not appear in this group is line radiation, a sharp spike or dip on the graph that represents a concentration or dearth of photons at a specific wavelength. Each chemical element has a unique pattern of emission lines (spikes) and absorption lines (dips), which astronomers find essential for calculating the composition, temperature, and density of an emitting object.

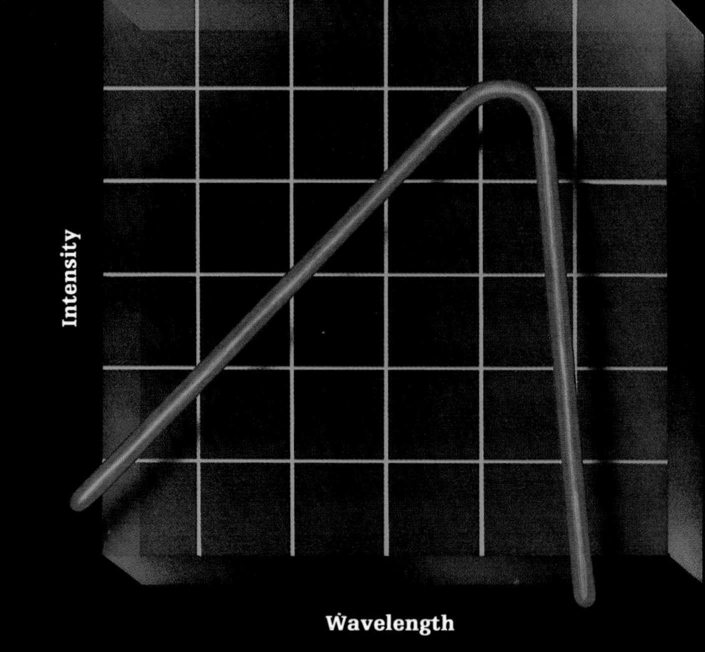

A peaked curve is the hallmark of blackbody radiation, so named because its source absorbs radiant energy, just as a black surface soaks up light. A blackbody source, however, reradiates the energy at wavelengths characteristic of its temperature. All relatively solid objects—including human beings and icebergs, as well as planets, stars, and clouds of opaque gas—emit blackbody radiation.

The absorbed energy increases the motion of atomic particles within the source, raising its temperature. As the particles interact, they lose some of their heat, which is released as radiation. Eventually, if the level of the incoming energy remains constant, the temperature of the body and the consequent radiation also stabilize, creating a condition called thermal equilibrium. A body in this state emits a steady stream of photons, most of them with wavelengths close to a specific value. The peak of the blackbody curve occurs at this characteristic wavelength, which is shorter for hotter objects.

The blackbody radiation from a star indicates the temperature of its surface, which absorbs and reradiates energy from the nuclear fires at its core. A very hot star emits large numbers of short-wavelength photons that show up as a high blackbody curve close to the right side of the spectrum. A cooler, less active star displays a lower peak, shifted toward the left side of the graph.

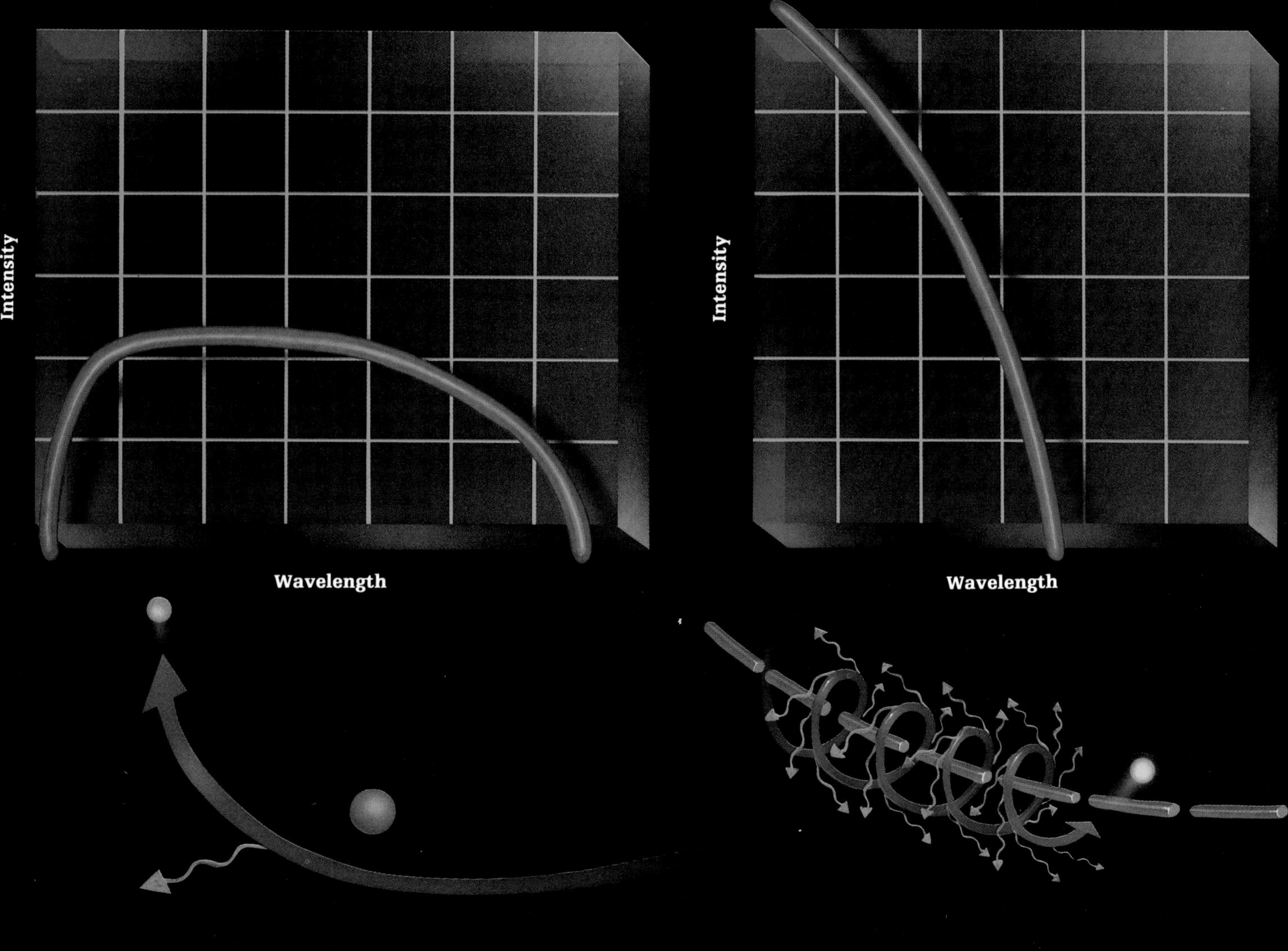

A curve shaped like a low hill is the sign of so-called free-free radiation, generated by the interaction of charged particles in thin clouds of ionized gas. When the path of a negatively charged electron *(blue dot in diagram)* is bent by its attraction to a positively charged ion *(red dot)*, the electron loses energy, released as a photon of radiation *(wavy line)*. The wavelength of any single photon is determined by two factors: the velocity of the electron that emits it, and the closeness of the electron's approach to the ion. Short-wavelength photons are always emitted by high-velocity electrons. The nearer an electron comes in passing an ion, the shorter the wavelength of its radiation. Because close approaches are least likely in extremely tenuous clouds, the wavelength at which a free-free curve levels off (called the turnover point) gives clues to the density of the cloud.

A sharply plunging curve betrays the presence of synchrotron radiation, emitted when an electron traveling near the speed of light encounters a strong magnetic field. The electron *(blue dot in diagram)* is pulled forward and around the line of magnetic force *(dashed purple line)* in a spiral path. Constantly changing course, the electron loses energy in the form of photons *(wavy lines)*, whose wavelengths depend upon the electron's velocity and the strength of the magnetic field. High velocity and a strong field produce a tight spiral that eventually saps the electron's energy; it may emit short-wavelength photons for hundreds of years, then emit at longer and longer wavelengths until it finally stops radiating altogether. Slower-moving electrons in a looser spiral may go on emitting long-wavelength photons for thousands of years. First discovered in experiments with synchrotron particle accelerators, this form of radiation from space often points back to dense, collapsing stars, which are known to have the requisite strong magnetic fields.

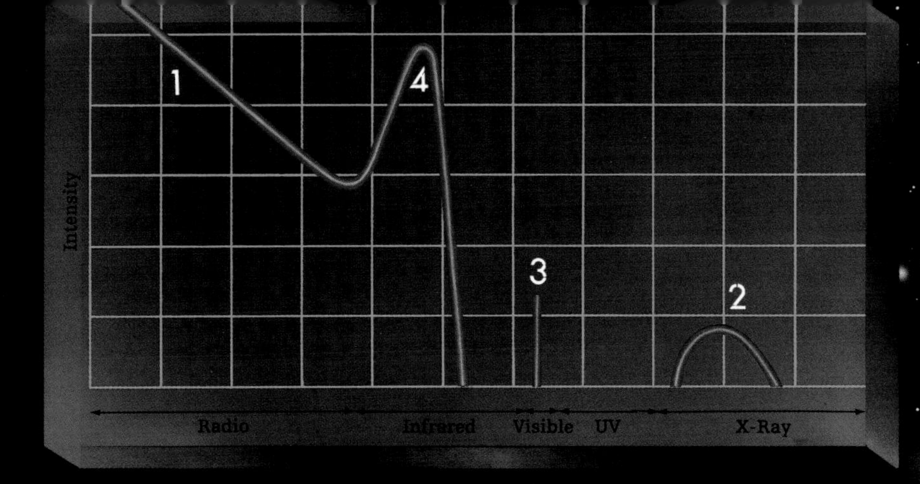

Message from a Supernova

The spectrum above contains four clues that point to features in a nearly invisible celestial object *(right)* —an object that carries in itself evidence of the long-ago and sudden end of a star.

One prominent hint is the sharply falling synchrotron radiation curve in the radio part of the spectrum *(1, above),* marking the presence of high-velocity electrons in a strong internal magnetic field *(1, right).* The turbulence of an exploding star often generates such a field, which may linger for as long as 100,000 years after the cataclysm.

A strong free-free element at x-ray wavelengths *(2)* indicates a cloud of thin, ionized gas at incredibly high temperatures, up to 40 million degrees Kelvin. One likely source of such heat is the violent collision of interstellar gas clouds, pushed by shock waves from some violent past event. The gas along the expanding shock front identifies itself as hydrogen (a principal stellar component) by emitting photons at specific wavelengths, seen here as an emission line *(3)* in visible red light.

The final piece of evidence is a blackbody curve in the infrared *(4),* a tip-off that the source includes warm dust regions separate from the extremely hot x-ray sources. The wide temperature range, strong magnetic field, and rapidly expanding shock wave all support the conclusion that this radiation outlines a supernova remnant, the residue of a massive exploding star.

1

2

3

4

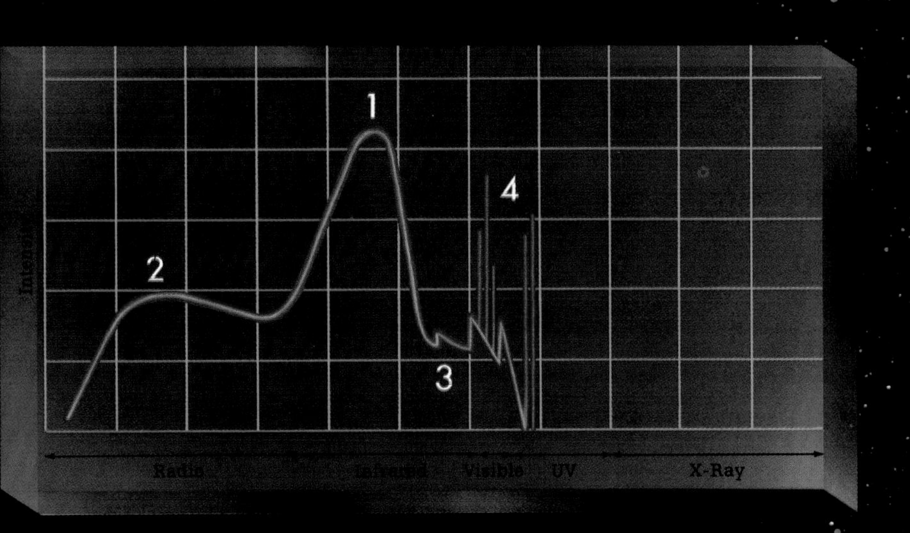

A White Dwarf's Nebulous Cloak

Hidden in the curves and spikes of the spectrum above is evidence of a slow celestial death. The position of the high blackbody peak in the infrared *(1)* suggests to an astronomer that the object contains a tenuous cloud of cool dust particles that absorb and reradiate energy from another source.

The slope in the radio portion of the spectrum *(2)* indicates radiation from free-flying electrons in a cloud of gas heated so it is ionized and thin enough to allow the electrons to remain unbound and radiating. The gas is ionized by a centrally located star, whose radiation heats the surrounding cloud to about 10,000 degrees Kelvin.

The saw-toothed curve *(3)* is created by visible light waves, also emanating from the ionized gas. When an electron comes too close to an ion within the cloud, the ion captures the electron, which gives off radiation that appears in the spectrum as a jagged peak. The spectral emission lines *(4)* indicate that the shell contains atoms of hydrogen, oxygen, carbon, and neon.

Taken together, the components of this spectrum reveal a thin, hot gas mixed with dust and glowing with energy from a small, hot star at its center. The description fits an object known as a planetary nebula, which is a slowly expanding shell of gas and dust blown away from a giant red star, the core of which has contracted to become a so-called white dwarf.

1

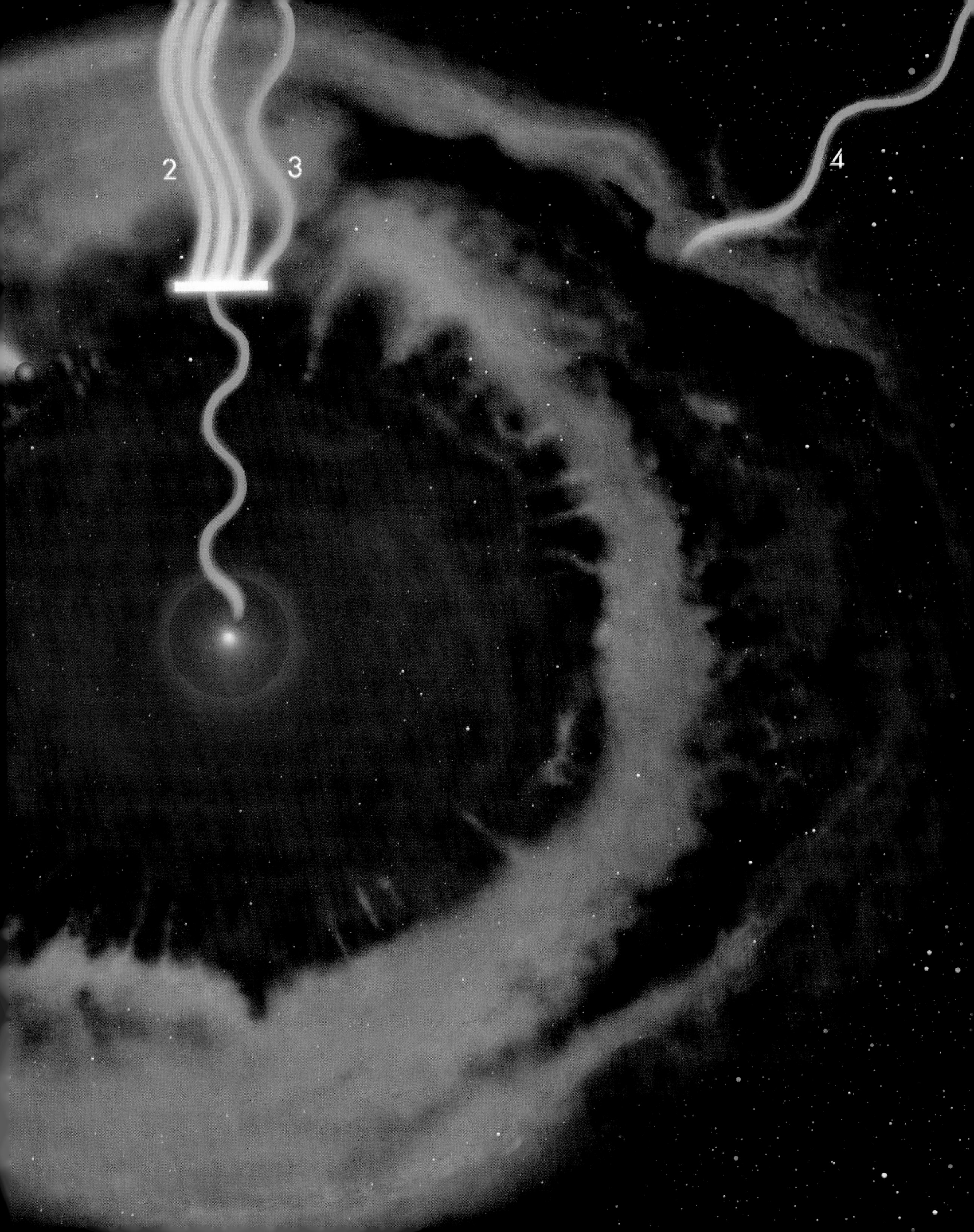
2
3
4

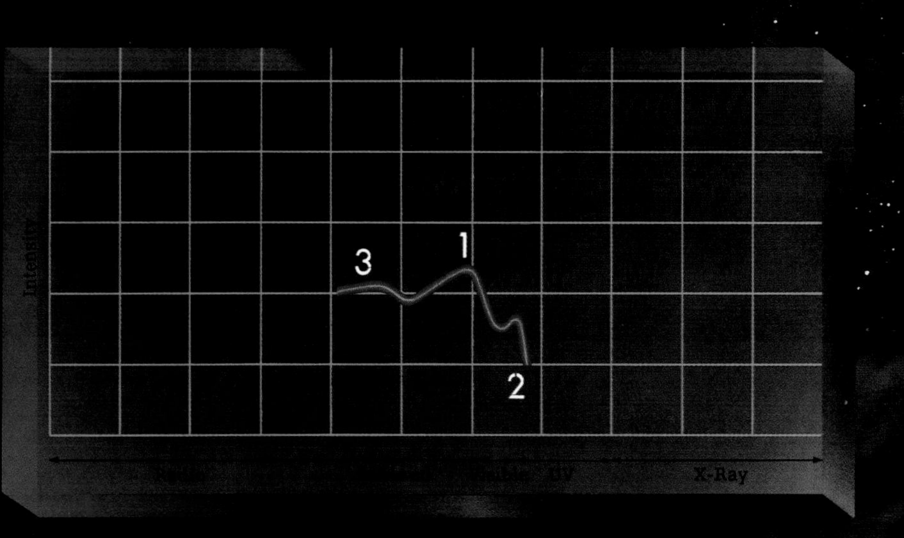

IDENTIFYING A SOLAR COUSIN

The spectrum above, seemingly simple in form, holds clues to the existence of a complex celestial entity. The blackbody curve peaking at visible-light wavelengths *(1)* shows that the object probably includes a star that is about as hot as the Sun—between 4,000 and 6,000 degrees Kelvin.

There is more to the puzzle, however. Instead of dropping off quickly on the right, as would a normal blackbody curve, this one extends into the ultraviolet *(2)*, signaling that some part of the source reaches temperatures of 10,000 degrees or more. The large amount of radiation is unlikely to be caused by flares on the star. But such temperatures could be produced by countless violent impacts on the star by matter falling onto it. An unusually large amount of infrared energy *(3)* indicates a possible source for the falling matter. The infrared probably emanates from many small dust particles, heated by the star, that radiate as individual blackbodies. Their temperature, varying with their distance from the star, ranges from 100 to 2,000 degrees.

Because this extended dust cloud is close to the star but does not obscure its light, astronomers conclude that it must take the form of a disk, one of the identifying features of a T Tauri star. Similar in mass and temperature to the Sun but much younger, a T Tauri star slowly draws the contents of its dust-filled disk down to its surface, where the matter vaporizes on impact, releasing ultraviolet waves.

2
3

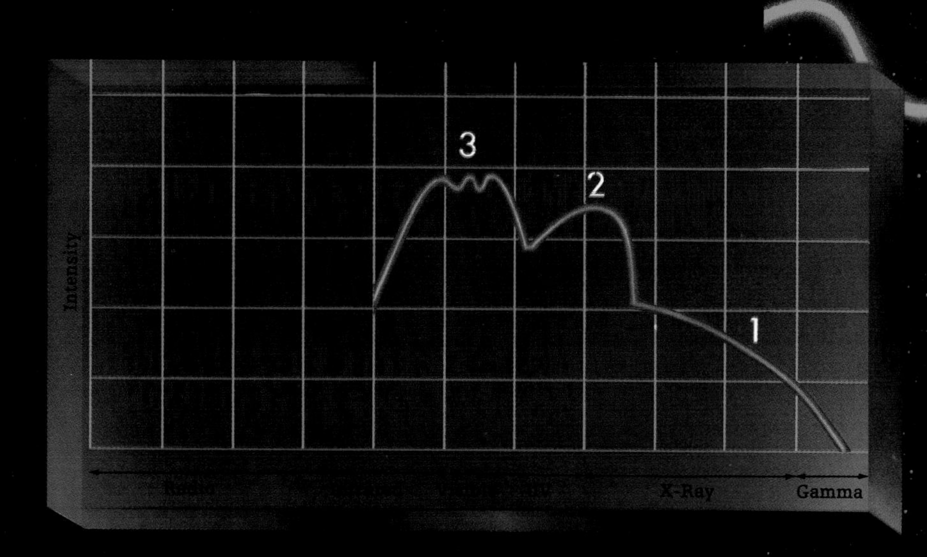

DECIPHERING A PULSING PUZZLE

This spectrum belongs to a little-understood object that gives astronomers an additional clue not evident from the graph: Its radiation pulses on and off over a period of three hours. The shape of the curve in the x-ray region *(1)* signals free-free radiation, produced by electrons moving through very thin ionized gas at unusually high temperatures. A blackbody peak in the ultraviolet *(2)* indicates an opaque object at 300,000 to 500,000 degrees. A second blackbody curve in visible light *(3)* is topped by serrations characteristic of cyclotron radiation, a less energetic cousin of synchrotron radiation that is also emitted by particles in a magnetic field. Together they point to ionized gas in thermal equilibrium at 10,000 degrees, associated with a magnetic field.

One scenario *(right)* suggests a binary system with a small but massive white dwarf star and a much larger, reddish companion. Matter pulled from the large star by the white dwarf's gravity funnels down to one of the dwarf's magnetic poles, heating electrons, which collide with ions in the hot falling matter and release x-ray photons *(1)*. The ultraviolet blackbody radiation *(2)* comes from the star's own absorption and reradiation of energy from the falling matter. The optical radiation *(3)* originates in a halo around the pole: Hot electrons emit cyclotron radiation as they pass through the dwarf's magnetic field, and clouds of heated gas produce blackbody radiation. The pulses occur because the star rotates, turning its radiant polar region first toward Earth and then away.

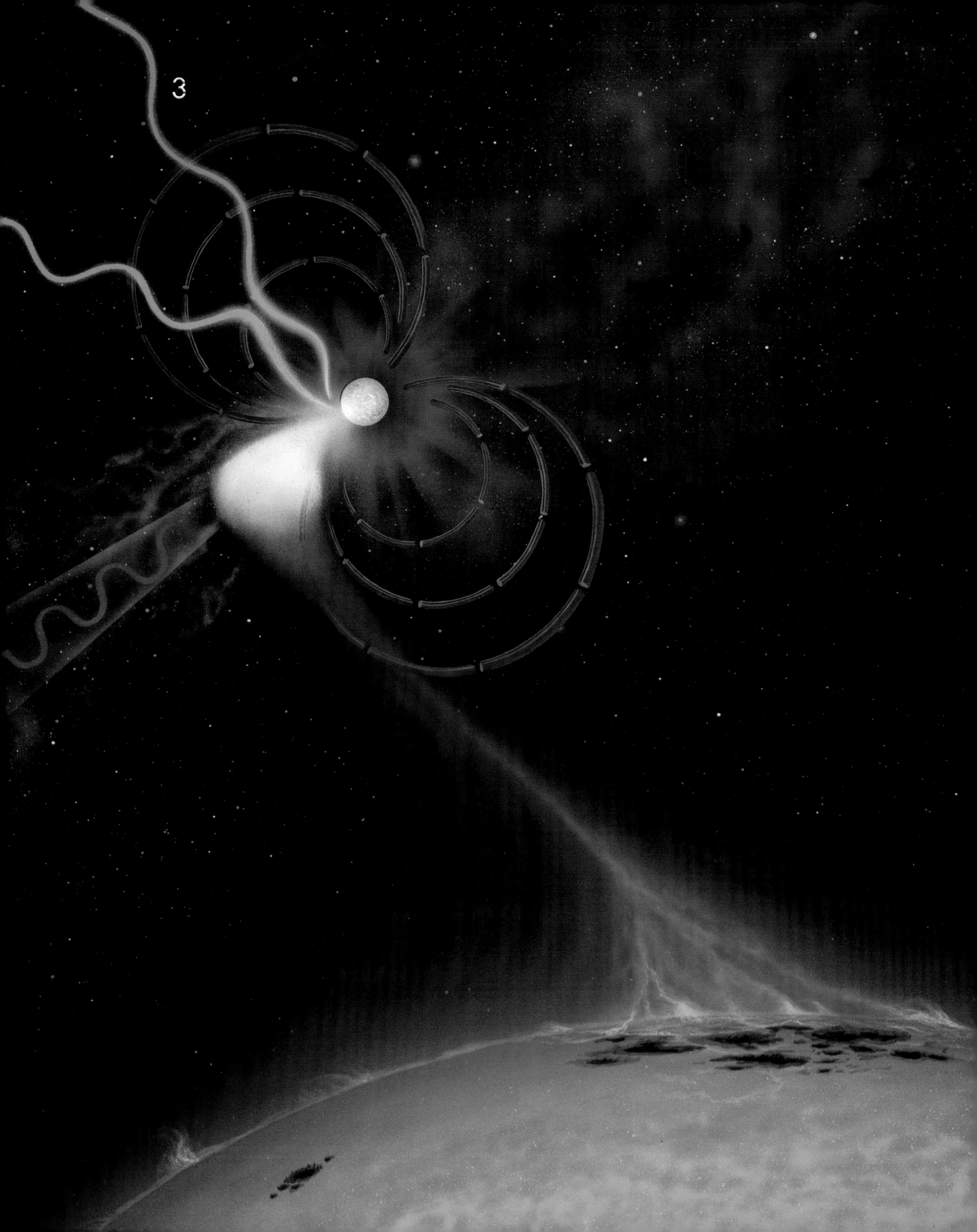
3

2/The Energetic Universe

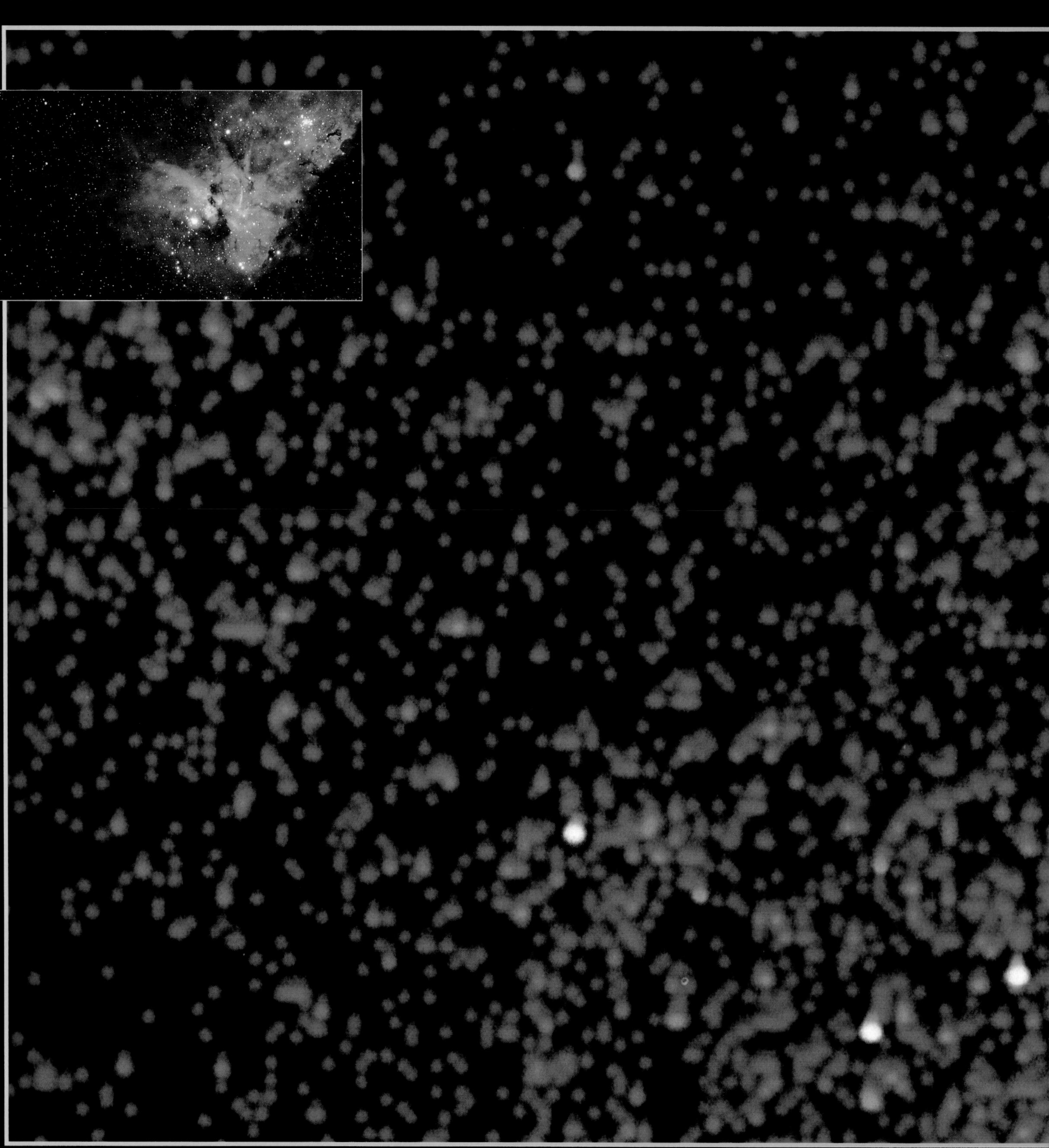

Hot, young stars dominate an x-ray image of the Carina nebula, a vast cloud of dust and gas 9,000 light-years from Earth. The stars stand out clearly in this picture; in an optical image *(inset),* their light is almost lost in the glow of the cloud.

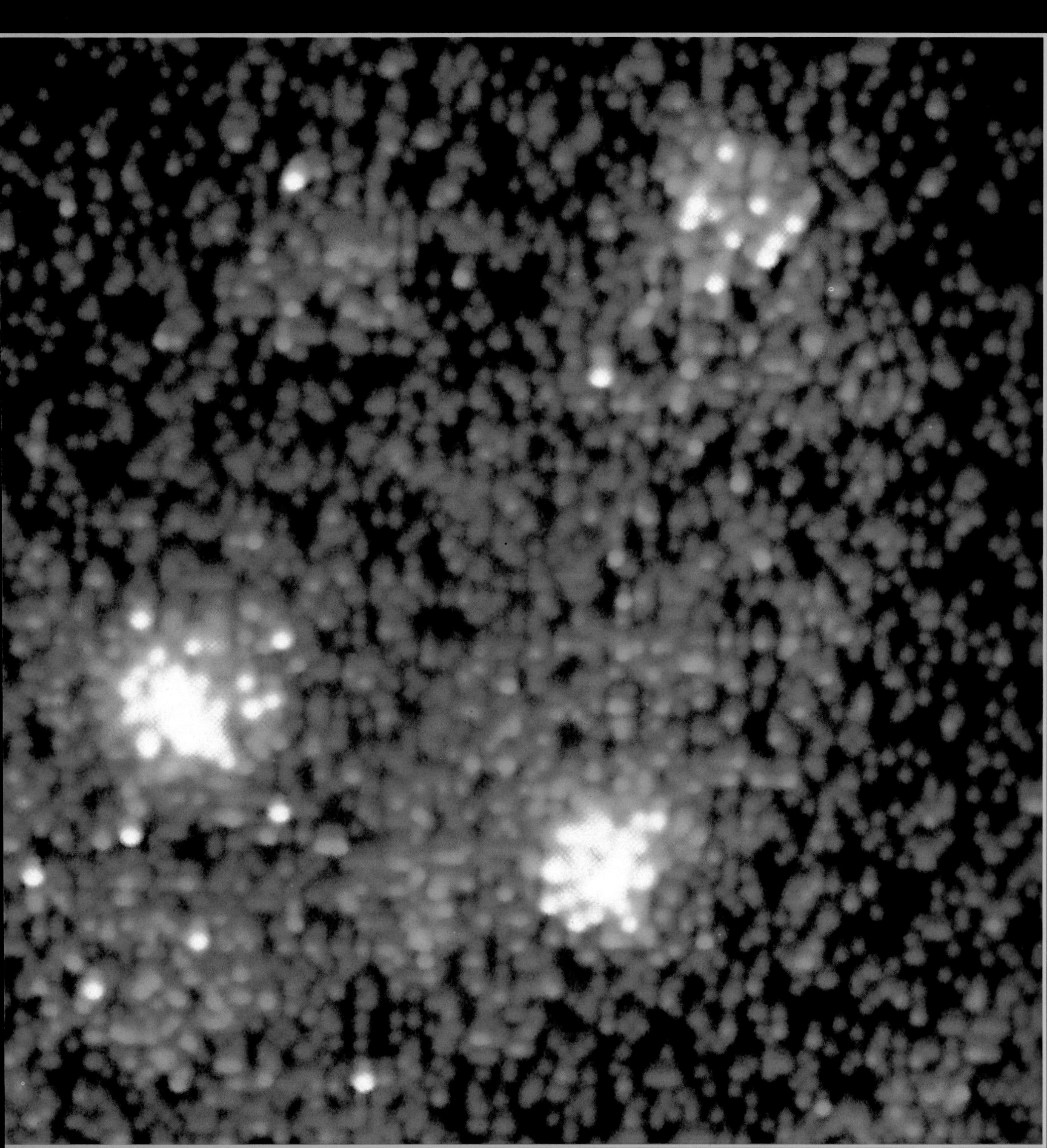

he desert sky blazed with stars on the night of June 18, 1962, and the Moon was just one day past full, but the astronomers gathered at White Sands Air Force Base in New Mexico ignored the burnished light above them. The sky they looked for was not the one they saw; their goal was the x-ray universe, a realm invisible to human eyes and inaccessible to earthbound instruments. At one minute to midnight, a svelte Aerobee rocket arced over the base, bound for the upper atmosphere. On board were a trio of instruments called proportional counters, primed to sense x-radiation from space and to gauge its strength. Essentially modified Geiger counters, the devices were sealed boxes filled with a gas, usually argon, and strung with one or more high-voltage wires, or electrodes. When x-rays penetrated the box, they would collide with argon atoms and knock loose a shower of electrons; the total charge of the electrons was proportional to the energy of the x-rays that precipitated it (hence, "proportional" counter). The electrons would then trigger an electrical current, which was amplified to give a measure of the rays' energy.

Twice before, the waiting scientists had been frustrated in their attempts to boost such equipment above the atmosphere, a gaseous shield that absorbs all incoming x-rays at altitudes between fifteen and sixty miles. One rocket had exploded during launch a year before, and a second sent back no data because the doors over its instrument panel had stuck shut. But the team had persisted, partly due to the stubbornness of its leader, an aggressive young astrophysicist named Riccardo Giacconi.

Giacconi was the head of the space research division of an independent Boston contractor, American Science and Engineering, a position he came to after entering astronomy almost by accident. During his youth in Genoa, his true love had been architecture, and his graduate career in the physics lab at the University of Milan had borne little relation to space science. Nevertheless, Giacconi went on to study particle physics at the University of Indiana and then Princeton. As a research associate at Princeton, his attentions were directed to the collisions of cosmic rays—high-speed particles from deep space—with atoms in Earth's atmosphere. (Cosmic rays are so named because mystified scientists once thought they were high-energy waves of radiation, similar to gamma rays.) Upon leaving school in 1959, his interest in cosmic rays led him to join forces with Bruno Rossi, another Italian immigrant, who taught at the Massachusetts Institute of Technology.

Rossi had pioneered the study of cosmic radiation in the 1940s, and by 1959 had become the principal consultant to AS & E. Intrigued by the potential of space probes for exploring the nonvisible, high-energy end of the electromagnetic spectrum, he encouraged Giacconi to join the firm and begin an x-ray program. The younger man acted on the suggestion with characteristic zeal and began a campaign to win backing for a project to discover celestial sources of x-radiation.

Backers were not so easily found. The prevailing wisdom about astronomical x-ray sources was that they were few and feeble. Next to gamma rays, x-rays are the most energetic manifestation of electromagnetism, consisting of waves ranging in length from .01 to 10 nanometers. (One nanometer is one-billionth of a meter.) Those with longer wavelengths exhibit wavelike properties, but shorter ones behave more like particles, or photons. As a consequence, scientists classify all x-rays not by wavelength but by the energy each photon contains, which can vary from about 2,000 to 100,000 electron volts, or 2 to 100 kiloelectron volts (KeV). (By comparison, photons of visible light typically carry 2 electron volts, or eV.) Only the hottest or most energetic processes in the universe will produce such high-frequency radiation.

In the 1950s, American experimenters had sent instruments into the upper atmosphere atop surplus German V-2 rockets and detected x-rays emanating from the extraordinarily hot gases seething in the Sun's corona. However, these emissions were pitifully weak, a millionth of the Sun's total optical output. If x-rays from Earth's nearest stellar neighbor were only barely detectable, scientists reasoned, such radiation from distant stars would likely be lost across the vast stretches of space. An x-ray map of the sky would probably reveal a scattering of faint sources at best, especially since the sensitivity of the proportional counters was minimal.

After many rejections, Giacconi finally succeeded in winning partial backing for his plan from a classified Air Force laboratory, which agreed to finance a suborbital flight to search for x-radiation coming from the Moon. (Some theorists believed that energetic particles streaming constantly from the Sun might emit high-frequency waves as they bombarded the lunar surface.) Giacconi was also hoping that a look at the sky in this region would reveal objects that might be scarcely visible at optical wavelengths but would shine brightly in the x-ray range.

Now, much to the thrill of the team at White Sands, the instruments aloft in the Aerobee began to register x-rays as soon as the rocket punched through the screening atmosphere. For a brief five minutes before the rocket dropped back to Earth, the recording mechanism jittered, etching a distinct peak at regular intervals on the unrolling strip charts. To the team's surprise, however, the radiation did not seem to come from the Moon after all. Over the next few months the scientists analyzed and reanalyzed the data. After ruling out other possibilities, including equipment malfunction and contamination from ultraviolet rays, they concluded that they had instead found a very strong x-ray source in the southern sky within the constellation Scorpius. Although they named the object Sco X-1, they could determine nothing further about its nature.

Giacconi's announcement of the first x-ray source to be discovered outside the Solar System, which he made at summer's end to an international gathering at Stanford University, prompted incredulous colleagues to press for confirmatory readings. These came quickly within the next year from two additional AS & E flights as well as from one undertaken by the Naval Research Laboratory in Washington, D.C. Perhaps more intriguing than the detection of Sco X-1 was a second kind of signal picked up on the flights indicating that the universe was bathed in diffuse x-radiation. This background wash of radiation was at first attributed to the energetic particles that are trapped by the Earth's magnetic field in two rings known as the Van Allen belts, hundreds to thousands of miles above the planet. But the emissions were so uniform across the sky that scientists eventually dismissed this notion in favor of the view that the radiation was probably left over from the formation of the universe—a hypothesis that particularly fascinated cosmologists trying to marshal evidence for competing theories on the origins of space and time.

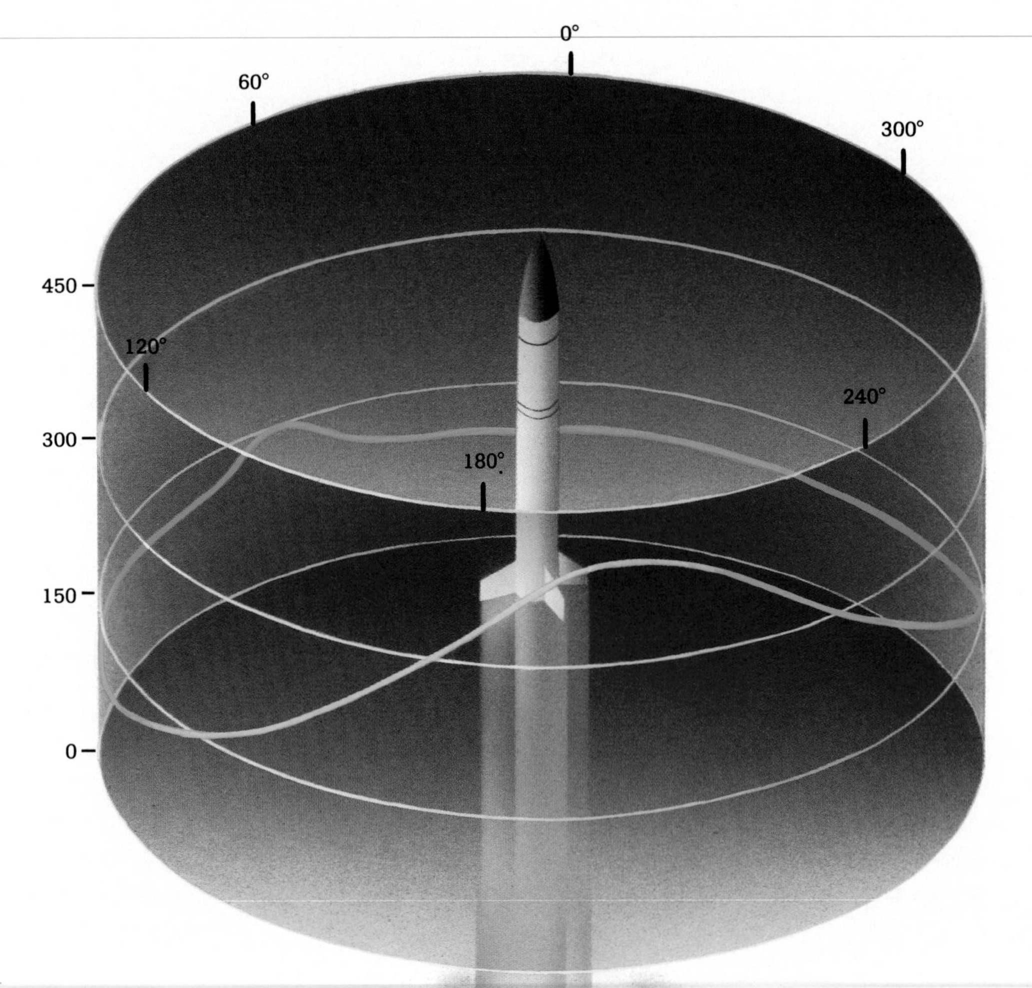

The rocket-borne experiment that detected the first cosmic x-ray source in June 1962 used an ingenious technique to survey the sky through 360 degrees. The Aerobee rocket, carrying three fixed x-ray detectors, turned at a rate of two revolutions per second while soaring above the atmosphere. As the detectors scanned the heavens, they recorded an increase in the number of x-ray photons each time they pointed at the constellation Scorpius, on a bearing of about 195 degrees *(peak on chart, left)*. A weaker signal appeared at 60 degrees, in the constellation Cygnus. But the counts never dropped to zero, indicating a background of x-radiation pervading interstellar space.

As Giacconi's experiment confirmed, rockets and spacecraft afforded astronomers hitherto unavailable opportunities to explore the universe beyond the ultraviolet range of the electromagnetic spectrum. For the first time, scientists could penetrate the atmosphere to gather high-frequency, high-energy radiation and unravel the message that it carried. Balloons and rockets could also take readings on fast-moving cosmic rays, which rarely reach the Earth's surface. During the next decade, researchers swept up in a spirit of derring-do pushed for expensive space missions on the chance that they would encounter, as one astronomer put it, "something really exciting." In the realm of high energies, the universe began to seem far more dynamic and violent than when viewed at optical wavelengths.

THE X-RAY FRONTIER

The success of the American Science and Engineering team spawned dozens of high-altitude balloon flights in the last years of the 1960s. Because the balloons could not rise entirely above the obscuring atmosphere, their findings were less satisfactory than those obtained with rockets, but they were still able to detect and record the most energetic sources of x-radiation. Soon, several groups began planning suborbital rocket forays. However, besides the investigators at AS & E, only the Naval Research Laboratory team, led by the astronomer Herbert Friedman, possessed the skill and luck both to design the innovative sensing equipment needed and to deal with the vagaries of rocket science, which was still a hit-or-miss affair.

In order to find out how the Sun might affect radio transmission, the diligent, self-effacing Friedman had carried out extensive research on solar x-rays during the 1950s, firing proportional counters into the upper atmosphere atop V-2 and Aerobee rockets. In the middle of the decade, he extended his studies to search for celestial x-radiation in the night sky. Purely by chance, he missed discovering Sco X-1, but it was his team's follow-up flights that gave scientific imprimatur to the work conducted by the less-experienced AS & E group.

After confirming the AS & E sighting, Friedman and NRL began a comprehensive survey of the sky, just as radio astronomers had done in the infancy of their discipline. In the process, the laboratory developed more-advanced detectors, so-called beryllium window counters, which were actually a collection of proportional counters grouped under a thin, credit-card-size sheet of the metal beryllium. The beryllium acted as a sieve for radiation, and a honeycomb of baffles, known as a collimator, restricted the field of view for incoming rays. These counters were ten times as sensitive as those riding on the first AS & E flight.

Meanwhile, Giacconi's team at American Science and Engineering also fabricated high-resolution collimators that enabled them, during a March 1966 rocket shot, to zero in on Sco X-1. Soon a group of Japanese researchers linked the mystery object to a faint blue star, an identification quickly verified by researchers operating the 200-inch optical telescope at Palomar Observ-

REVELATIONS OF GRAVITATIONAL LENSES

Gravitational lens systems—natural telescopes stretching billions of light-years through the cosmos—give astronomers a unique perspective on remote phenomena. A gravitational lens is a massive body, such as a galaxy or cluster of galaxies, interposed between Earth and a distant object, often a quasar. The lens's gravitational field bends and magnifies the object's light, creating a bright image for observers on Earth. Although a perfect alignment of Earth, the lensing body, and the distant entity would result in a ring-shaped image, such a precise arrangement is unlikely. Since the first lens system was discovered in 1979, no lensed image has shown the ideal shape; often a gravitational lens produces two or more distorted images, usually spots or arcs.

In a perfectly aligned gravitational lens system, radiation from a remote quasar *(far right)* speeds past an intervening galaxy toward Earth. The center galaxy's gravitational field bends the radiation equally in all directions, focusing it so that an earthbound observer, looking along the dotted green line of sight, sees a bright ring, apparently at the distance of the light source. This illusion is called an Einstein ring, for Albert Einstein, who developed the hypothesis of gravitational lensing from his general theory of relativity.

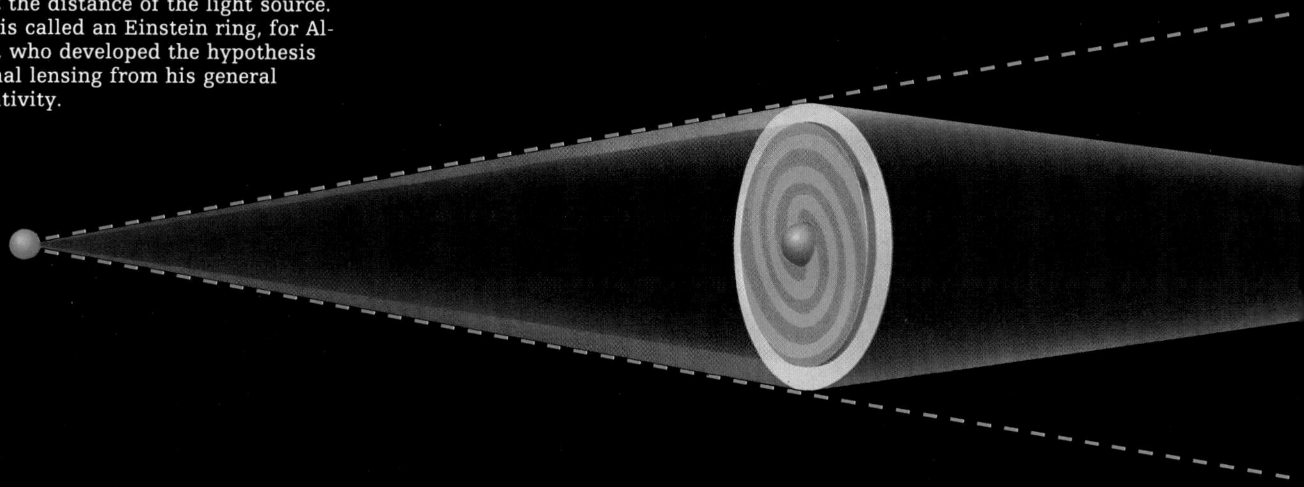

Unlike a glass lens, which bends different wavelengths of light unequally, a gravitational lens imparts exactly the same curve to all electromagnetic radiation. Therefore astronomers can study lensed objects not only in visible light but at radio and other wavelengths as well. With the data gleaned from gravitational lens systems, astronomers can draw inferences about the radiation source and the lens itself, finding answers to important questions about the nature of the universe. One concerns the amount of so-called dark matter, estimated by theorists to account for more than 90 percent of the total mass of the universe. As its name suggests, this form of matter emits no discernible radiation, so it can be detected solely by its gravitational effects. By comparing the visible mass of the lens to the mass required to produce the observed bending of radiation, astronomers can calculate how much dark matter the lens contains.

Another benefit is improved estimates of the distance to remote quasars, crucial to understanding the size, age, and expansion rate of the universe. When a lensed quasar flares, one of its two images brightens before the other, because its radiation is less bent and thus travels a slightly shorter path. Because radiation travels at a constant speed, the time interval between the two flares can be converted into a precise measure of the difference in path length. When this value is inserted into the deduced geometry of the system, scientists can compute the distance that each set of light waves travels from the quasar.

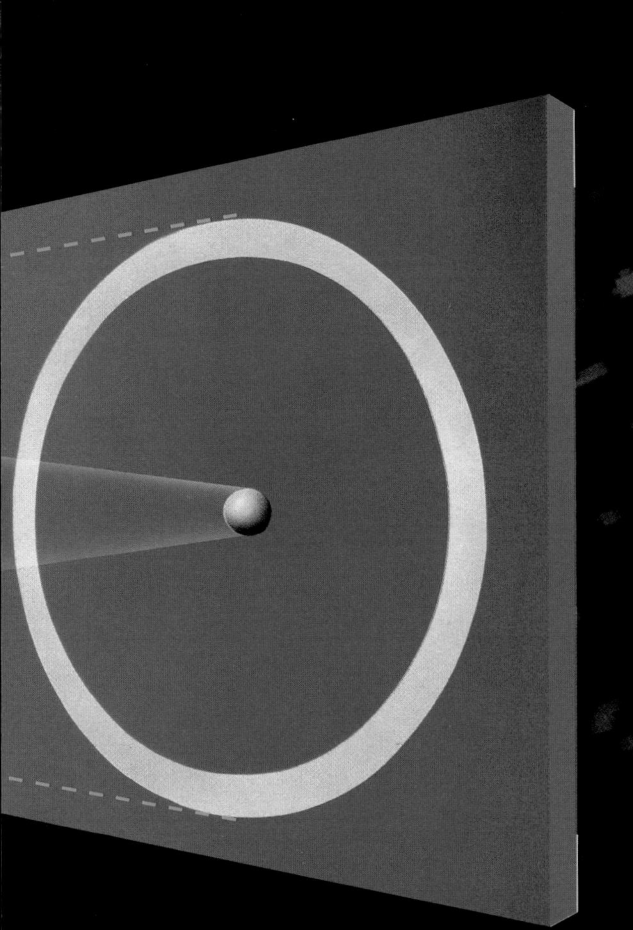

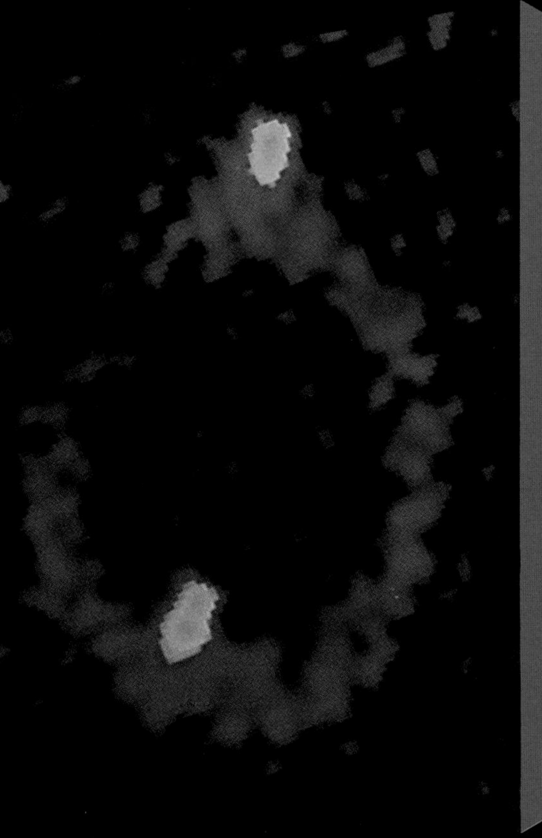

Radio waves emitted by a so-called core jet in a distant galaxy, shown here in false color, mark what is believed to be the first sighting of an Einstein ring. Recorded at the Very Large Array radio telescope by Princeton astronomer Jacqueline Hewitt and colleagues, the ring is slightly flawed, perhaps due to uneven bending of the radio waves by an asymmetrical distribution of mass in the lens galaxy; another possible cause is a slight misalignment of the lensing system.

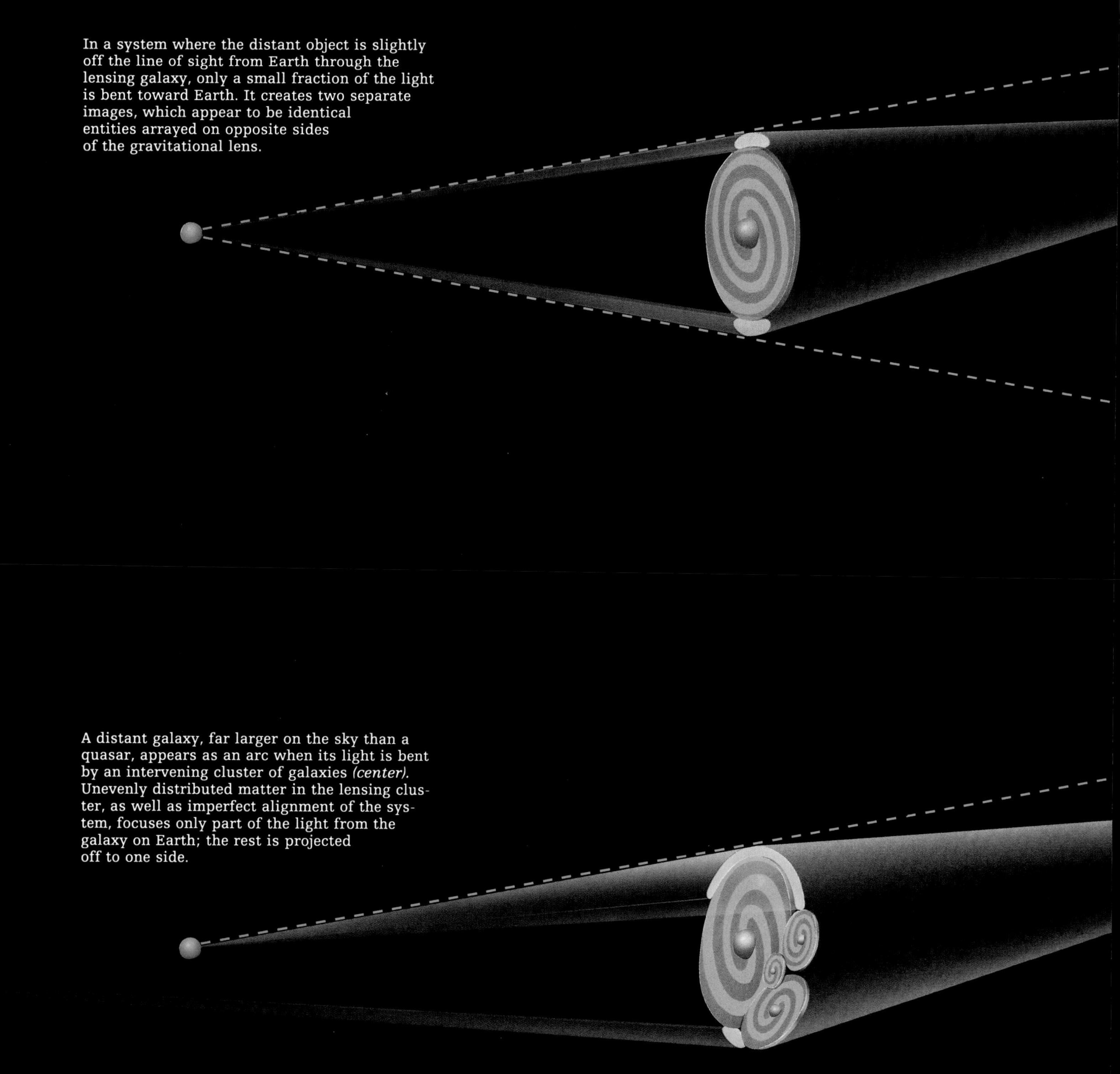

In a system where the distant object is slightly off the line of sight from Earth through the lensing galaxy, only a small fraction of the light is bent toward Earth. It creates two separate images, which appear to be identical entities arrayed on opposite sides of the gravitational lens.

A distant galaxy, far larger on the sky than a quasar, appears as an arc when its light is bent by an intervening cluster of galaxies *(center)*. Unevenly distributed matter in the lensing cluster, as well as imperfect alignment of the system, focuses only part of the light from the galaxy on Earth; the rest is projected off to one side.

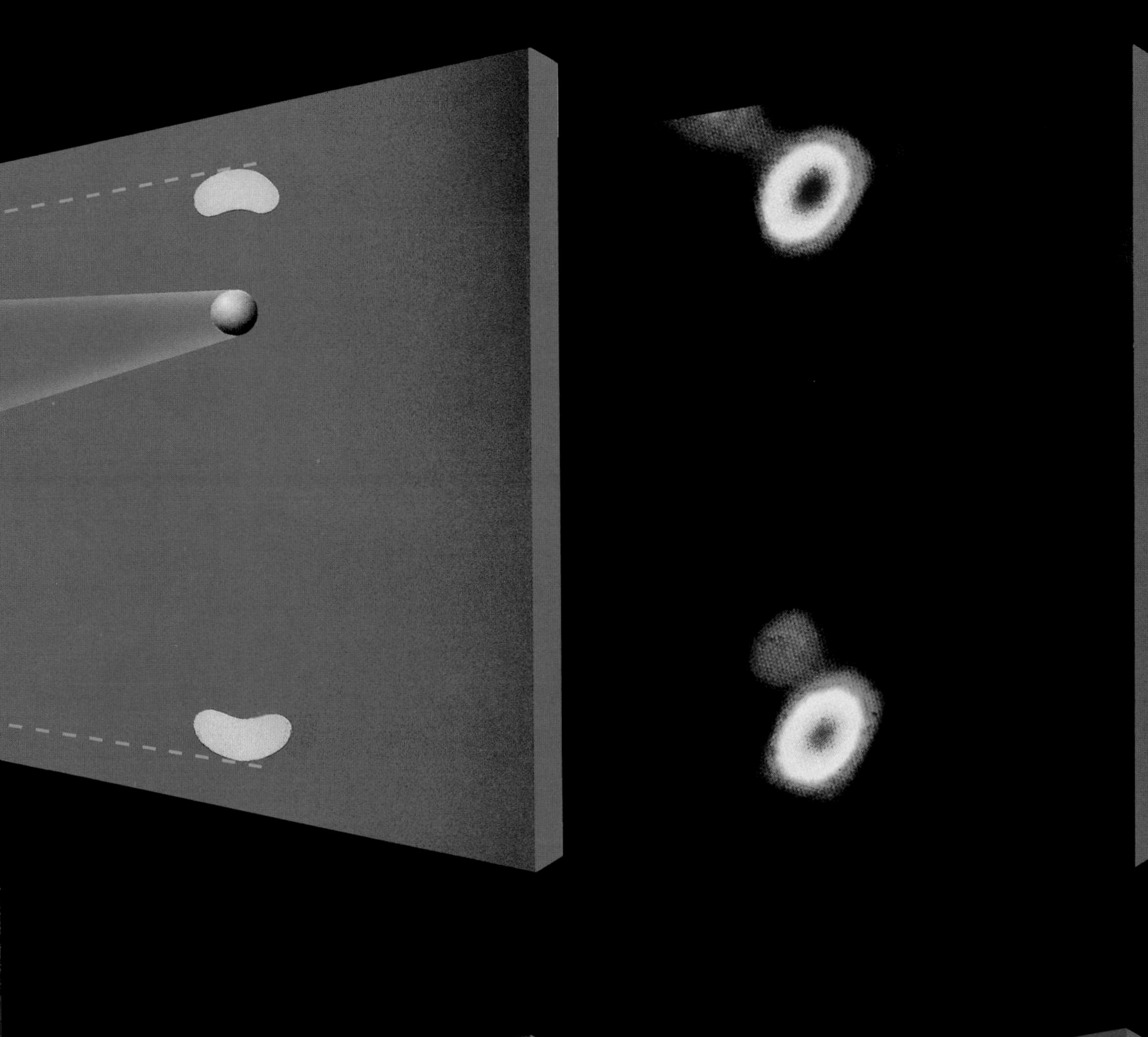

This false-color radio image records the first discovery of a complex gravitational lens system, in 1979. Scientists initially thought the red and yellow images at top and bottom were twin quasars, but they later determined the images were two views of a single quasar. Because the system is imperfectly aligned, the lower image appears much closer to the lens galaxy.

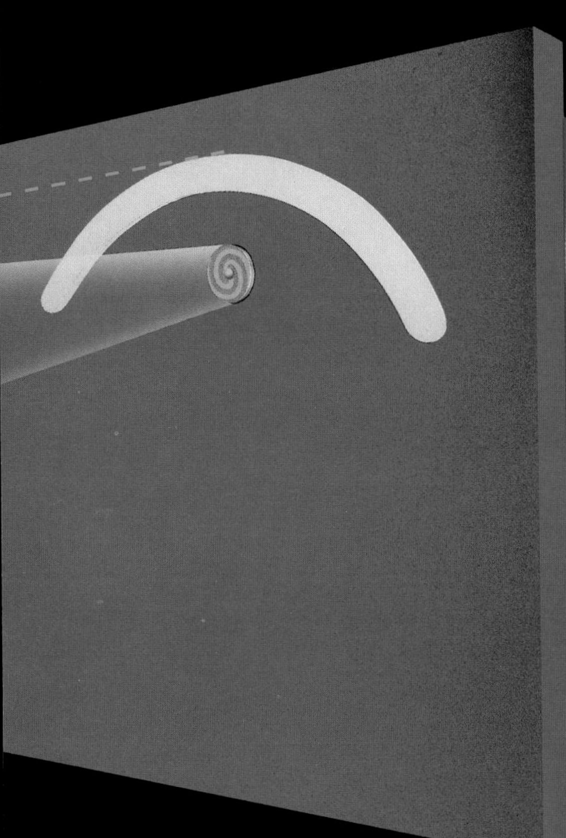

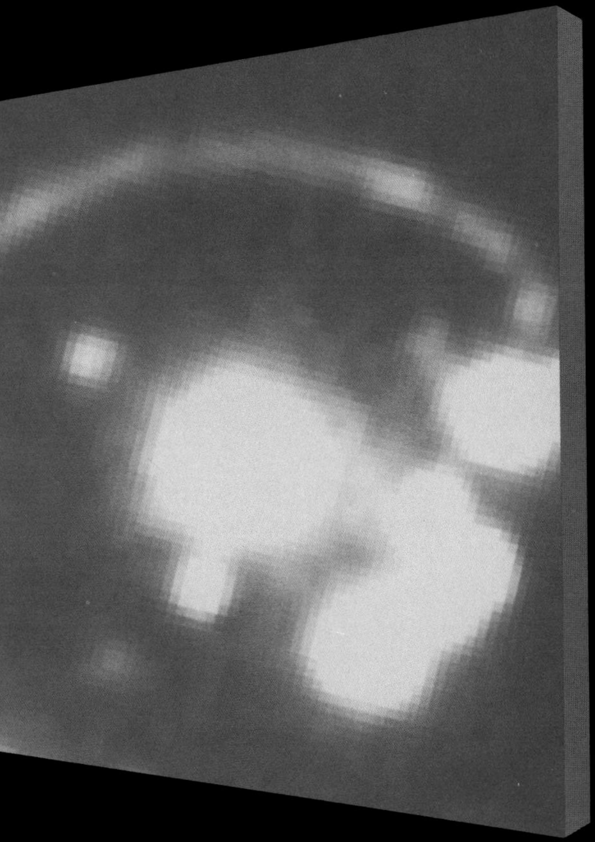

A great luminous arc stretches over the image of a galactic cluster. The arc is actually light from a distant galaxy, magnified and brightened by the gravitational field of the cluster. Astronomers have concluded that to produce such a large, bright lensed image, the cluster must contain a substantial amount of dark matter.

atory north of San Diego. Even so, astronomers were hard put to formulate a coherent picture of the forces involved in the generation of celestial x-radiation. The x-ray output of the source so dwarfed that of the Sun that scientists could not imagine what processes might come into play. If the Sun's dim x-radiation resulted primarily from collisions of ionized atoms and electrons in its hot outer shell of gases, something more was clearly transpiring among the stars.

Perhaps, surmised researchers, they had finally located one of the most exotic members of astronomy's theoretical bestiary: a neutron star. The existence of these small, extremely dense stars had been predicted in the 1930s by astronomers Fritz Zwicky of the California Institute of Technology and Walter Baade of Mount Wilson Observatory. According to their radical theory—which was widely doubted at the time—when a massive star blows apart in a supernova explosion, its core can collapse into a spinning mass of neutrons while its outer shell goes whirling into space. The x-ray astronomers thought that the neutron star might have a surface temperature in the millions of degrees, hot enough to emit steady x-radiation.

Soon after the Sco X-1 sighting, while casting about for possible explanations, Friedman revived the neutron star hypothesis to account for another x-ray emitter in the Crab nebula, the wispy vestige of a spectacular and widely chronicled supernova explosion that took place in 1054 AD. And yet, much to Friedman's disappointment, an experiment he performed late in 1964 produced an ambiguous reading, neither confirming nor denying his hypothesis. Three years later, however, Friedman's link between neutron stars and x-rays was vindicated when Cambridge University astronomers Antony Hewish and Jocelyn Bell discovered a strange radio signal from an object they called a

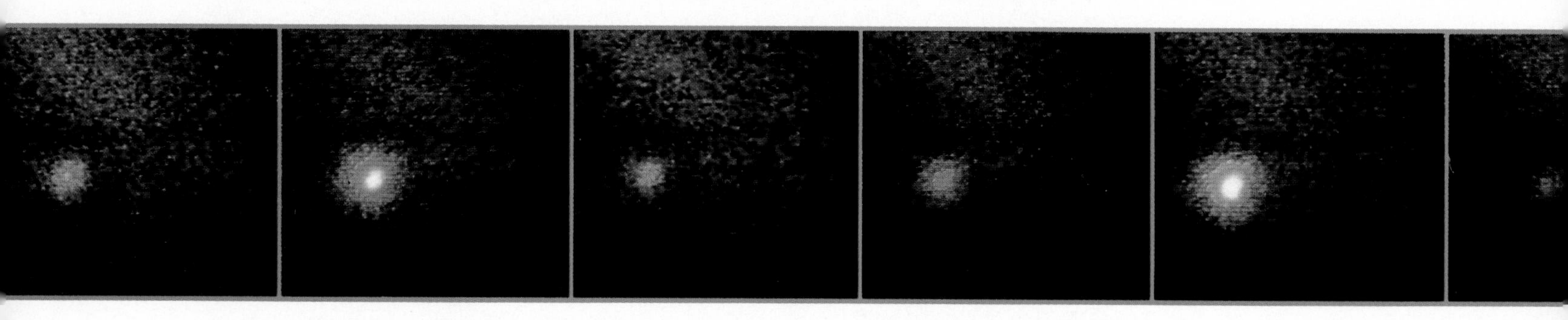

pulsar. The best explanation for the signal's regular bursts of radiation was that the source was a rapidly spinning neutron star. The object apparently produced radiation not from a superheated surface but by a mechanism known as synchrotron emission, in which high-energy particles spiraling along the force lines of a strong magnetic field give off radiation at various wavelengths along two magnetic poles. The radiation would reach observers on Earth only when the rotating beam passed over the planet's surface. Spurred on by the discovery, Friedman decided to look again at the Crab

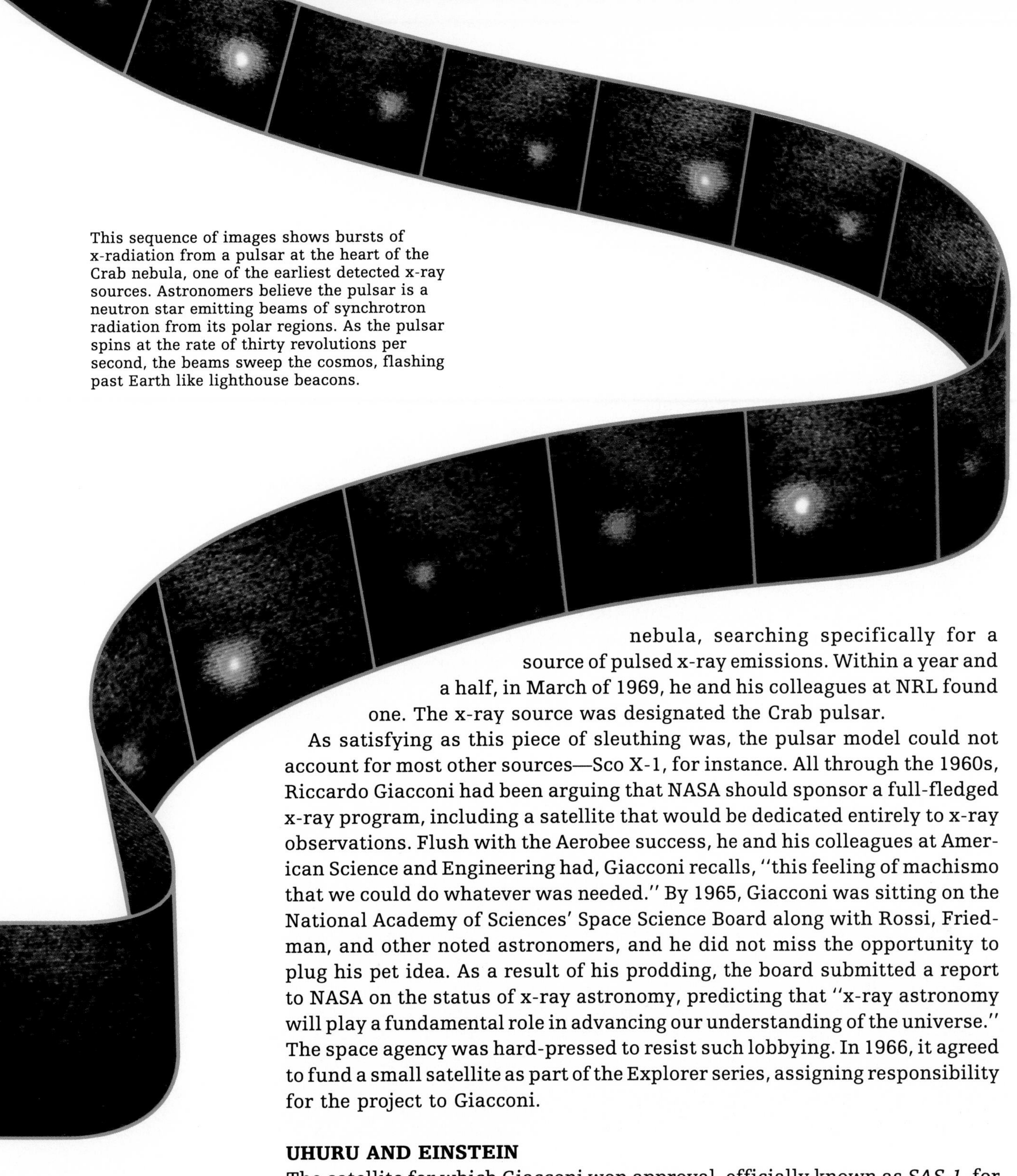

This sequence of images shows bursts of x-radiation from a pulsar at the heart of the Crab nebula, one of the earliest detected x-ray sources. Astronomers believe the pulsar is a neutron star emitting beams of synchrotron radiation from its polar regions. As the pulsar spins at the rate of thirty revolutions per second, the beams sweep the cosmos, flashing past Earth like lighthouse beacons.

nebula, searching specifically for a source of pulsed x-ray emissions. Within a year and a half, in March of 1969, he and his colleagues at NRL found one. The x-ray source was designated the Crab pulsar.

As satisfying as this piece of sleuthing was, the pulsar model could not account for most other sources—Sco X-1, for instance. All through the 1960s, Riccardo Giacconi had been arguing that NASA should sponsor a full-fledged x-ray program, including a satellite that would be dedicated entirely to x-ray observations. Flush with the Aerobee success, he and his colleagues at American Science and Engineering had, Giacconi recalls, "this feeling of machismo that we could do whatever was needed." By 1965, Giacconi was sitting on the National Academy of Sciences' Space Science Board along with Rossi, Friedman, and other noted astronomers, and he did not miss the opportunity to plug his pet idea. As a result of his prodding, the board submitted a report to NASA on the status of x-ray astronomy, predicting that "x-ray astronomy will play a fundamental role in advancing our understanding of the universe." The space agency was hard-pressed to resist such lobbying. In 1966, it agreed to fund a small satellite as part of the Explorer series, assigning responsibility for the project to Giacconi.

UHURU AND EINSTEIN

The satellite for which Giacconi won approval, officially known as *SAS-1,* for Small Astronomy Satellite 1, proved an incalculable boon to the study of high-energy phenomena. By 1970, Giacconi, Friedman, and others had unveiled more than thirty notable x-ray sources within the Milky Way galaxy, in addition to a handful outside it. However, they knew little more than that the sources existed and that some might be the recently discovered pulsars. *SAS-1* was their first chance to take a prolonged look at the x-ray sky.

The 141-pound satellite blasted off from a seaborne platform in the Indian

Ocean along the coast of Kenya on December 12, 1970—coincidentally, Kenyan independence day. Exuberant scientists promptly renamed the orbiter *Uhuru,* the Swahili word for "freedom." For two years—before falling back into the atmosphere and burning up—*Uhuru* orbited around the equator, where its equipment would be relatively untouched by interference from charged particles circulating in the Van Allen belts. The satellite confirmed astronomers' suspicions that the universe harbored far more violence than visible light would indicate. *Uhuru's* instruments located 339 x-ray sources throughout the sky; most of the brightest sources lay along the plane of the Milky Way, suggesting that they were part of this galaxy.

With *Uhuru's* detectors aimed into space, astronomers could begin to assemble a library of vital statistics, giving theorists the raw material from which to build an understanding of the dynamics of x-ray generation. The research was not without surprises. It became apparent, for instance, that many sources varied in power over time. Some seemed to turn on and off with great precision; others regularly fluctuated in intensity. The pulses from the source Centaurus X-3, for example, arrived once every 4.8 seconds and ceased altogether every 2.1 days. The intensity of another source, Cygnus X-1, fluttered randomly.

Puzzling out these bizarre behaviors, theorists added a new mechanism, called binary accretion, to the list of processes known to produce x-radiation. Many sources seemed to be identified with the bright stellar bursts known as supernovae. Astronomers had decided that most supernovae occur when one star in a binary system—which consists of two stars orbiting a common point

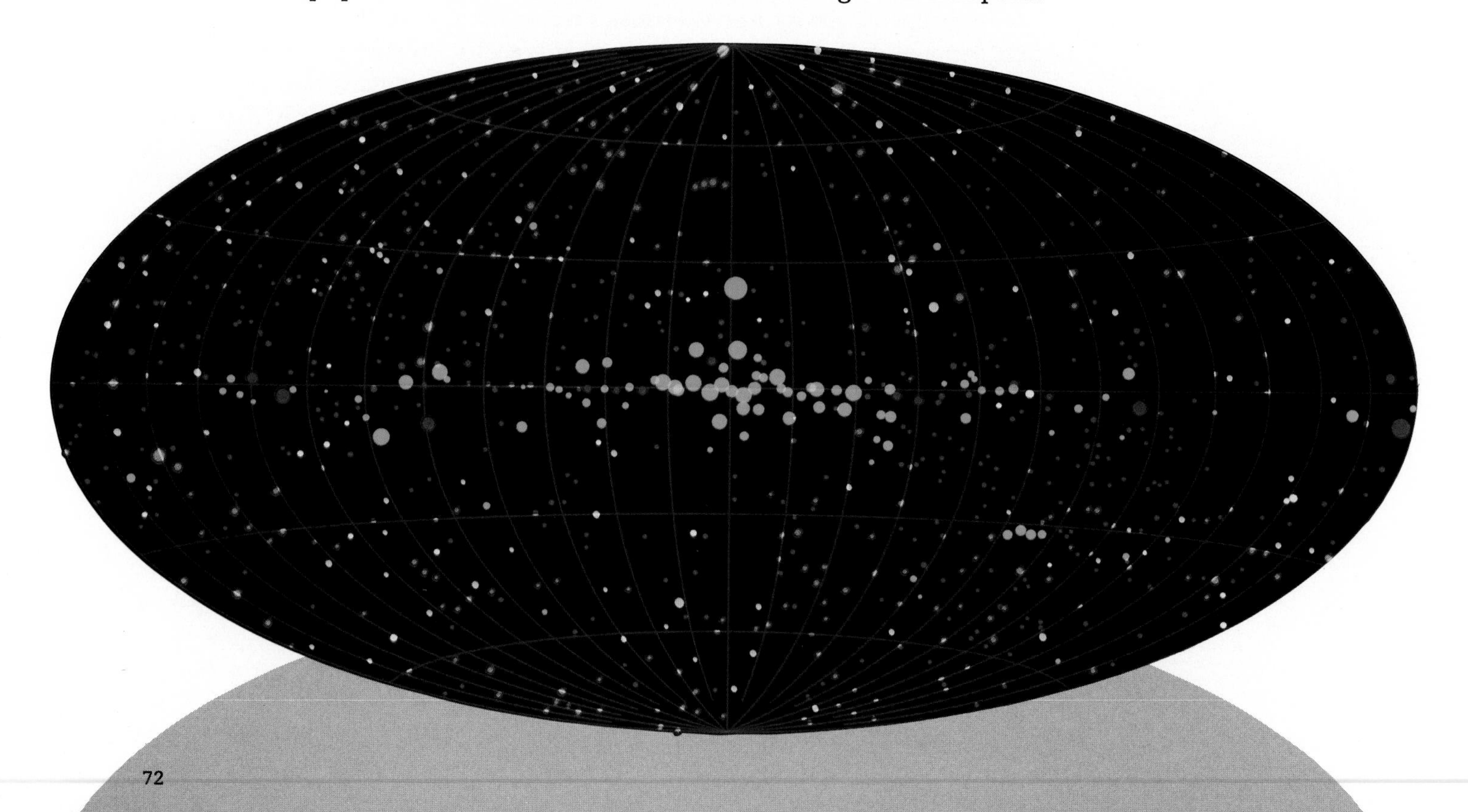

of mass—pulls hydrogen off its less-massive companion. When the accumulated gas reaches a critical mass on the first star, the star first collapses, then explodes. In some cases, the more-massive star might collect so much material that it collapses into a neutron star. Could it be, wondered astronomers, that some x-ray sources were the results of such supernovae: binary systems in which one partner was a neutron star? If Centaurus X-3 was part of such a system, in which both stars revolved around a mutual center of gravity, it should periodically disappear—its radiation blocked as it was eclipsed by the companion—and then reappear when the source emerged from the other side. This neatly fit the behavior of Centaurus X-3.

Researchers also found that the pulses from Centaurus X-3 underwent a cycle of modulation known as a Doppler shift, varying slightly in wavelength just as they should if a star were moving steadily away from, and then toward, a viewer on Earth. Further investigation revealed that the pulses themselves were linked to the star's period of rotation and were being produced at the poles. Apparently, as the immense gravity of the neutron star tugged matter away from its companion, the gas flowed toward the neutron star along the lines of its magnetic field, concentrating most heavily at the poles. Accelerating toward the star's surface, the atoms gave up their energy in the form of x-rays.

The Centaurus explanation in turn appeared to fit Sco X-1, the first x-ray star. In 1967, noted Soviet astronomer Iosef Shklovsky had suggested that the blue star seen at Sco X-1 had a neutron star partner, a massive object that pulled the first star's gas around it like a superheated shroud, generating the high-energy radiation. By the mid-1970s, other astronomers had confirmed his suspicions: Sco X-1 exhibited a periodic variation in brightness that matched a binary model.

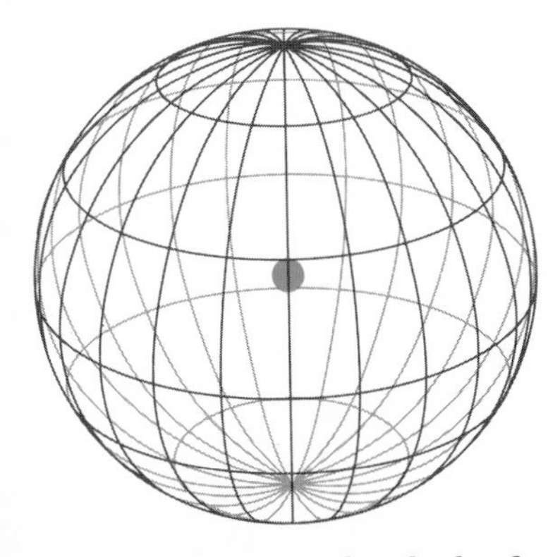

The map at left shows hundreds of x-ray sources detected by *HEAO-1*, the first High Energy Astronomy Observatory satellite, during a seventeen-month survey of the celestial sphere *(above, Earth at center)*. Sized in logarithmic proportion to their intensity, they include supernova remnants *(blue)*, galaxy clusters *(pink)*, normal stars *(white)*, binary stars *(green)*, galaxies and quasars *(yellow)*, and unidentified sources *(red)*. *HEAO-1*, launched in 1977, not only increased the catalog of known x-ray objects from 350 to 1,500 but also discovered a new black hole and several new classes of x-ray-emitting stars.

The binary theory helped astronomers understand an even more peculiar case, the mysterious source at Cygnus X-1. The visible star, a blue supergiant, appeared to be circling an unseen companion once every 5.6 days. Calculating the mass of the invisible star from its giant partner's movement, astronomers came to the astonishing conclusion that the smaller object should be at least six times as massive as the Sun—far too weighty even for a neutron star. The only possibility left was that the companion was a black hole, the infinitely dense product of a collapsed star. The source's odd, fluttering emissions might have been produced as matter siphoned off the companion star fell toward the black hole and heated up to the point where it released x-radiation.

Having discovered that the x-ray galaxy was dominated by binary sources, astronomers remained largely ignorant of the details: What was the spectrum of these emitters? How and where were x-rays produced in more commonplace stars? Once again, Friedman and Giacconi took up the cause of the x-ray community. Working with NASA, Friedman and other scientists designed the first in a series of satellites, the High Energy Astronomy Observatory 1 *(HEAO-1)*, a device that produced an important survey of the x-ray sky between 1977 and 1979, as well as performing valuable experiments in the gamma ray spectrum. The indefatigable Giacconi championed an even more ambitious program on *HEAO-2,* scheduled for launch in 1978. His creation would be armed with an x-ray telescope that represented a thousandfold improvement in sensitivity over earlier instruments.

Conceived by Giacconi and Rossi in 1960, the telescope was a modification of an x-ray microscope invented by a German physicist, Hans Wolter. To focus incoming radiation, Giacconi nested two cylindrical mirrors inside an aluminum tube. The mirrors were shallow curves, one parabolic, one hyperbolic. Instead of passing through the mirrors' highly polished, gold-coated surfaces, x-rays would graze along them and gently deflect to a common focal point *(pages 102-103).*

With *HEAO-2,* which was launched from Kennedy Space Center in November 1978 and dubbed the Einstein Observatory in honor of the great physicist's centenary, x-ray astronomers would glean images comparable to those in the optical field. The satellite's four detectors sent back word of x-ray objects a thousand times fainter than those spotted by *Uhuru.* As principal investigator, Giacconi, who had recently moved his lab to the Harvard Smithsonian Center for Astronomy, strove to ensure that as many researchers as possible got observing time on this, the first of the big, multi-institutional astronomical facilities.

Over its two-and-a-half-year lifetime, *Einstein* relayed the news of thousands of new x-ray sources in choice detail. By studying spectra, astronomers learned that the burgeoning menagerie included many hot, massive, recently formed stars. To their surprise, they also spotted a number of cooler stars, red dwarfs that had no coronas of hot gases from which to generate x-radiation. Flares, like those that erupt over sunspots, might perhaps have generated the x-rays in these cases; alternately, the radiation might have emanated from anomalies in the stars' magnetic fields. X-rays also streamed from the Milky Way's vast clouds of gas and dust—the residue of long-dead stars and the birthing ground of new ones.

Einstein went dark in 1981, plunging into the atmosphere, but Giacconi and crew had firmly established the importance of continuing observations in the x-ray range, and plans were soon afoot for still-more-sophisticated satellite-based programs for the 1990s. One West German national project, called ROSAT (short for Roentgen Satellite, after Wilhelm Roentgen, who discovered x-radiation in 1895), will carry a Wolter-type telescope three times more powerful than the one that rode on *Einstein* and is expected to net more than 100,000 x-ray sources, of which a tenth will be scrutinized in detail. The United States will boost ROSAT into space on the shuttle in exchange for a portion of the observing time. On its own, the U.S. will install another permanent satellite, the Advanced X-Ray Astrophysics Facility, in 1995. The *AXAF* will orbit for as long as fifteen years and deploy a telescope 100 times more sensitive than *Einstein*'s. By collecting information about such cosmic mysteries as quasars, black holes, dark matter, and exotic particles, *AXAF*

An Atlas-Centaur rocket blasts skyward from the Kennedy Space Center, carrying the Einstein Observatory, a satellite that could produce x-ray images comparable in quality to the best optical and radio pictures. Launched in 1978, the satellite transformed the study of non-solar x-radiation. Its data yielded detailed images of thousands of new sources, ranging from weak x-ray emitters in the Milky Way to quasars more than 10 billion light-years away.

may lead astronomers to a greater understanding of the size, age, and ultimate fate of the universe.

Even as x-ray astronomers gathered in their newfound wealth of data, other scientists hoped to duplicate their success with gamma rays. Discovered near the turn of the century, this radiation is the most potent in the electromagnetic gamut, occupying a surprisingly wide band of energies from .1 million to 10 million electron volts (megaelectron volts, or MeV) for relatively low-energy gamma photons to more than 10^{14} MeV for high-energy photons. At the high end, gamma rays carry more than 10 million trillion times the energy of visible light. Energy of this sort can appear only when matter is subjected to extraordinary temperatures, pressures, or speeds. In studying the far end of radiation, researchers explore the outer limits of physics, where, under unimaginably great forces, normal nuclear laws may not hold.

THE PURSUIT OF RARE PHOTONS

Gamma radiation is, in a sense, the smoke that points to underlying cosmic fires. Most gamma rays fall into the lower end of their range and are emitted as radioactive elements decay or as electrons interact with other matter. But a fraction belong to the high end of the spectrum: The higher the energy, the rarer the photon. Most of these photons appear to be the by-product of collisions between cosmic rays and other particles. Since the assorted cosmic particles yield gamma rays of varying energies, astronomers can, by examining the spectrum of received gamma rays, infer what mechanism produced them. For instance, electrons colliding with low-energy photons of starlight or passing through gas clouds yield photons below 50 MeV.

Like the x-ray universe, the gamma ray universe was virtually unknown before the advent of rocketry. In 1957, physicists William Kraushaar and George Clark of MIT had looked for nonterrestrial gamma rays using high-flying balloons but did not find any until they sent equipment up with the NASA satellite *Explorer 11* in 1961. Kraushaar was one of many experimental physicists who migrated into astronomy during the late 1950s and early 1960s, supplying instrument-building skills that most traditional sky gazers lacked. He had begun his career contemplating cosmic rays, like Giacconi, and was diverted into gamma ray studies by the theorist Philip Morrison of Cornell University, who argued that the best way to delve into the nature of cosmic rays was by indirection. In 1957, Kraushaar followed Morrison's advice and commenced the hunt for cosmic gamma rays, teaming up with Clark, a physicist who later worked with the American Science and Engineering x-ray group.

If Kraushaar was hoping for auspicious beginnings, he did not get them. His measuring device, called a scintillation counter, was basically a crystal of sodium iodide flanked by optical sensors capable of triggering an electronic pulse. When incoming gamma rays hit electrons in the crystal, they produced a flash of light, or scintillation; the sensors registered this and relayed a signal to the ground. The strength of the pulse corresponded to that of the incoming ray, and from the pulses it was possible to build a profile of the radiation. Since gamma rays were so scarce, however, this required protracted exposures. During the five months it was in operation, Kraushaar's equipment aboard *Explorer 11* registered a mere twenty-two celestial gamma photons, filtering them out from a distracting background buzz of gamma radiation created by cosmic rays hitting the upper atmosphere.

The meager results were indicative of things to come. Whereas the x-ray field had been buoyed by instant success, the gamma ray field would progress only by fits and starts, always in need of more data, more devoted researchers, more money for experiments, and technological miracles. Kraushaar himself, who is considered the father of gamma ray astronomy, was torn between it and the less frustrating pursuit of x-ray experimentation when he moved from MIT to the University of Wisconsin in 1965.

Still, Kraushaar's *Explorer 11* experiment did at least prove that with enough patience and time one could collect low-energy gamma rays. Kraushaar's next experiment, stowed aboard NASA's third Orbiting Solar Observatory for a year-long sojourn beginning in 1967, delivered further testimony to the worth of gamma ray studies. Because Kraushaar's wide-angle scintillation counter was searching only for gamma ray emission, other ordinarily bright objects—the Sun, the planets, and the stars in the Milky Way—faded to inconsequentiality. But the galaxy was not entirely dark. Instead, its central plane glowed, especially toward its midpoint, with high-energy gamma radiation—possibly generated from the collisions between cosmic rays and the hydrogen that suffuses interstellar space.

Throughout the 1970s, numerous gamma ray experiments, primarily operated from Department of Defense satellites, confirmed the radiation from the galactic plane, as well as locating other forms of gamma ray activity. The most perplexing of these were the gamma ray bursts, happened upon by a group of scientists led by Ray Klebesadel in New Mexico at the Los Alamos Scientific Laboratory (renamed the Los Alamos National Laboratory in 1981). Klebesadel and his colleagues were monitoring the sky for gamma ray output not from stars but from possible hydrogen bomb tests that would be in

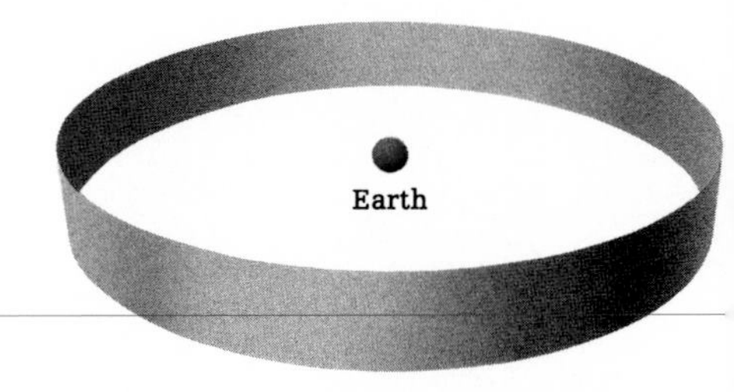

Produced by the European satellite *COS-B* from a survey of a narrow band of sky *(above)*, the gamma ray map of the Milky Way below reveals about thirty different objects. Because of *COS-B*'s low resolution, most of the sources are blurred; the powerful Vela pulsar appears as a white circle to the right of center. Gamma rays are exceedingly rare: Only one or two photons per hour arrived from even the brightest sources, compared to counts of a hundred photons commonly registered by x-ray detectors aboard *Einstein.*

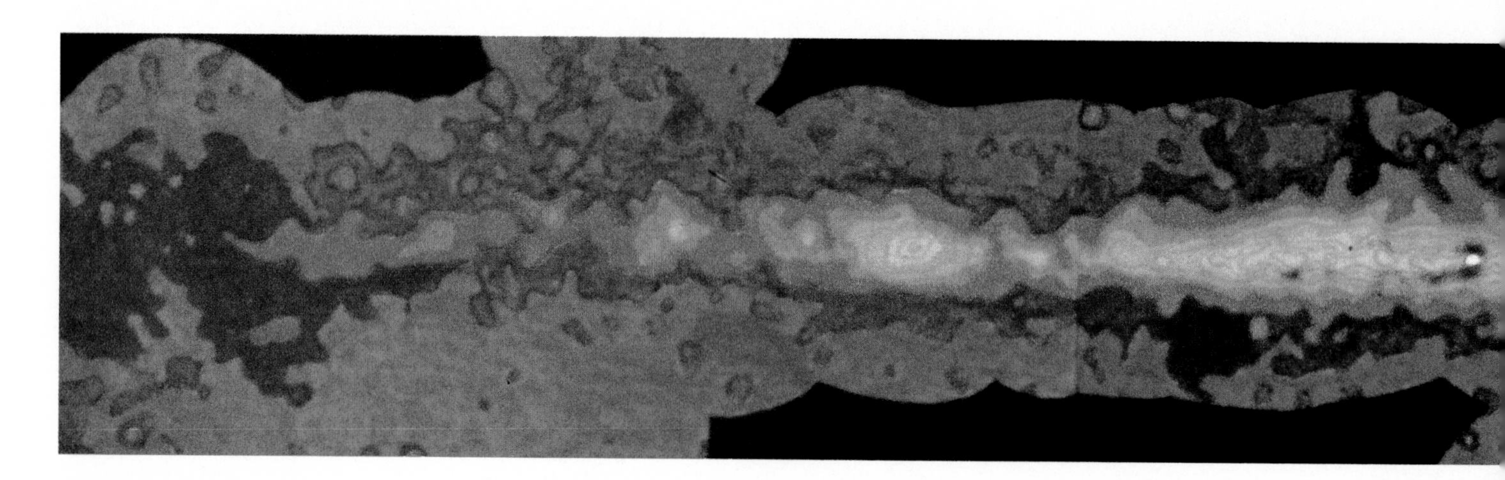

violation of the 1963 American-Soviet test ban treaty. In 1969, the team detected a burst of gamma radiation that could not be clearly associated with a bomb test. Between then and July 1972, they logged fifteen more such events, bursts of low-energy rays arriving at random from different areas of the sky. Each brief flux, though lasting half a minute or less, carried more than 100,000 times the power of the Sun. Thinking that the bursts might come from recent stellar explosions, the Los Alamos scientists searched their records for likely events and could find none that fit. At a loss to explain the phenomenon, they speculated that the surges might derive from a hypothetical type of supernova that did not radiate at visible wavelengths, or from a neutron star with an unthinkably intense magnetic field.

Gamma ray scientists acutely felt the limitations of their instruments. The central problem was that they could not determine the origin of a given packet of radiation with any certainty. Under the direction of Carl Fichtel, another convert from cosmic ray physics, researchers at the Goddard Space Flight Center forged an ingenious solution to the problem, using a variation on a device called a spark chamber. In a normal spark chamber, as in a scintillator, incoming gamma rays trigger an electrical signal. But Fichtel's instrument could, in addition, register the angle of a ray's entry. High-energy gamma rays entering the device would collide with an atom and knock free a pair of electrons; the charged particles would follow the trajectory of the original radiation and leave a trail of sparks along a series of metal plates. By projecting backward along the electrical paths, scientists could trace the gamma ray's celestial road. The invention was developed in time to make the launch of NASA's second Small Astronomy Satellite in November 1972. Seven months later, *SAS-2* suffered a power failure, but by then Fichtel's group had gathered enough data to map the broad contours of the high-energy gamma sky.

Soon astronomers were able to fill out the picture with information gleaned from a spark chamber aboard a European satellite called *COS-B,* which from 1975 to 1982 surveyed the Milky Way. Among other things, *COS-B* sensed a wash of feeble gamma radiation coming from outside the galaxy, and within one to two degrees helped to fix the locations of thirty discrete gamma ray

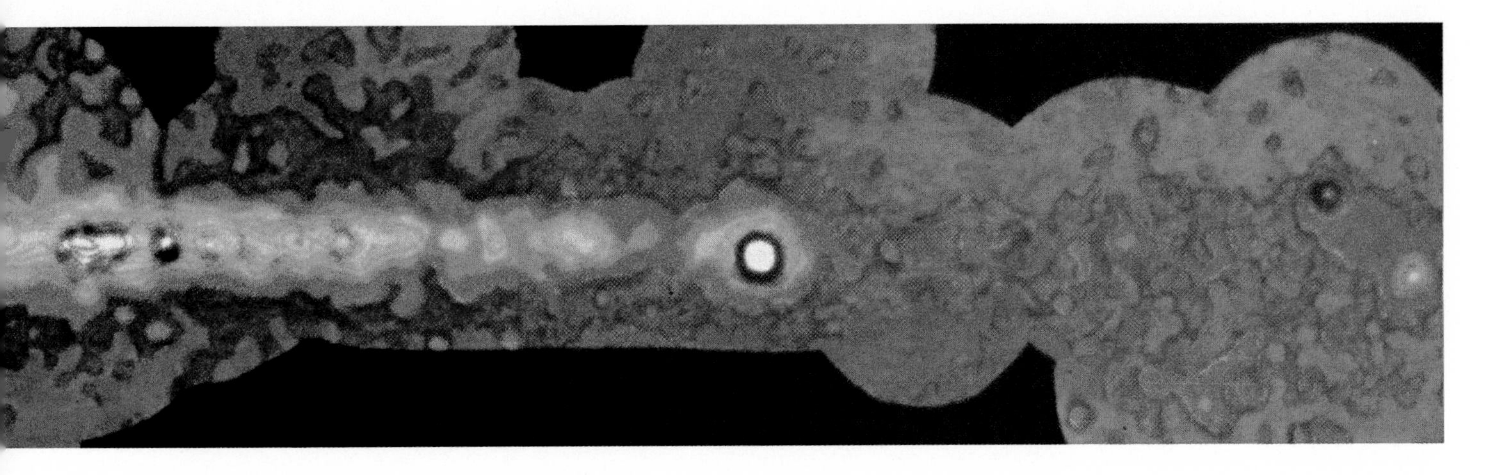

sources within the Milky Way. Although it was not possible to unequivocally match any one of them with a visible object, some almost certainly coincided with known sources of radio or x-ray energy. Among the most prominent of these were two x-ray sources, Cygnus X-3 and the Crab nebula, and a pulsar in the debris of the Vela supernova that radiated many times more energy at gamma than radio wavelengths. The existence of a gamma source among the celestial ruins in Vela indicated that some high-energy gamma radiation did indeed derive from huge stellar explosions, while the potent, fluctuating signals from Cygnus X-3 led astronomers to suspect that gamma rays and x-rays could be generated by the same mechanism: the flow of superheated gas from a companion star into a black hole.

Despite the contributions of satellites *SAS* and *COS-B*, astronomers were yearning to determine more precisely the dimensions and spectra of radiation sources. In 1979, these demands were partially met by an instrument invented by Allan Jacobson, one of the most imposing—and most unusual—figures in the gamma ray field.

THE BIRTH OF HEAVY ELEMENTS

No one would ever have marked Allan Jacobson for a physics whiz kid. Born in 1932, he struggled through the Chattanooga, Tennessee, public schools, ending high school with a C− average. After a stint in the Air Force, Jacobson tried to break into show business in Los Angeles. His impressive bass voice won him a contract from RCA, and he recorded an album of songs from the Rogers and Hammerstein musical *Pipe Dreams.* But the play went up in smoke on Broadway, and with it, Jacobson's singing career. In search of another line of work, Jacobson enrolled at the University of California, Los Angeles, on the GI Bill and drifted into physics, attracted by the visual beauty of spectrographs. Here he found a new vocation. In the late 1970s, he perfected a sensitive new spectrometer built around crystals of the semiconducting material germanium. By cooling crystals to −297 degrees Fahrenheit, he reduced the random "noise" usually generated by heat and was able to fine-tune signals from incoming gamma rays. The spectrometer vaulted into space aboard the third High Energy Astronomy Observatory in September 1979 and soon provided some clues to one of the earliest mysteries of gamma ray astronomy. The halo of gamma radiation observed toward the center of the Milky Way contained emissions that registered at 511 KeV, which scientists believe to be the energy associated with antimatter. The most likely source for the radiation: a supermassive black hole. Physicists theorize that within the extremely hot, dense masses of infalling matter near the black hole, gamma rays will run into electrons and create pairs of electrons and their antimatter counterparts, positrons.

Despite such results, *HEAO-3*'s new germanium detectors did not eliminate the inherent difficulty of gamma studies. It took Jacobson and a team of specialists two years to sift through the *HEAO-3* data, striving to separate true signals from the spurious. But the effort was worthwhile. Finally confident

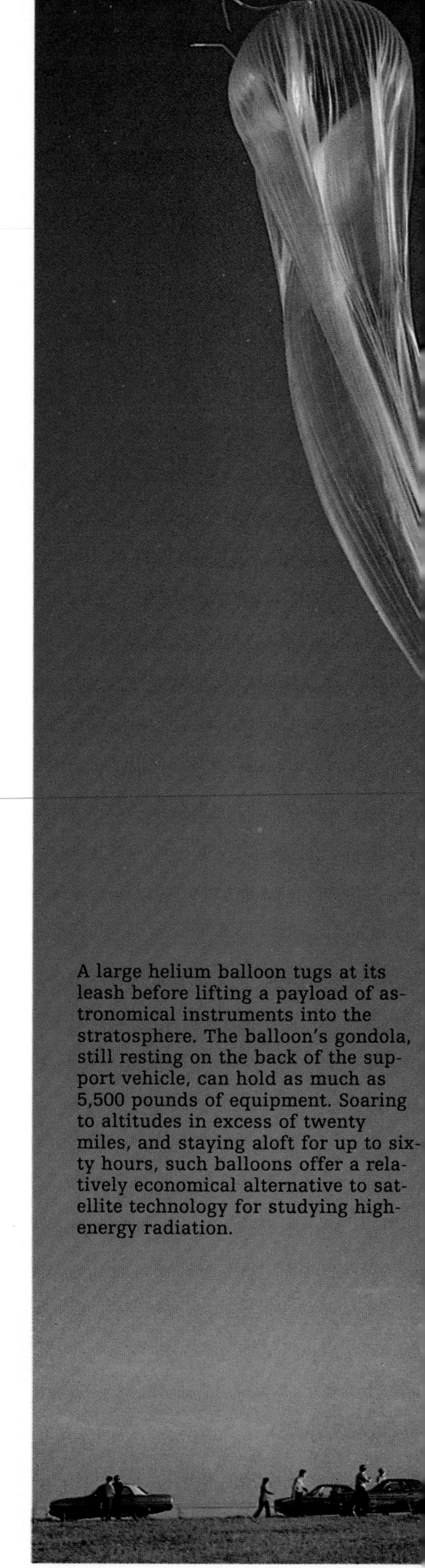

A large helium balloon tugs at its leash before lifting a payload of astronomical instruments into the stratosphere. The balloon's gondola, still resting on the back of the support vehicle, can hold as much as 5,500 pounds of equipment. Soaring to altitudes in excess of twenty miles, and staying aloft for up to sixty hours, such balloons offer a relatively economical alternative to satellite technology for studying high-energy radiation.

that they had found a bona fide message, the researchers announced another major discovery in the November 1984 *Astrophysical Journal.* They had apparently captured the characteristic emissions of a decaying radioactive isotope of aluminum in low-energy gamma rays coming from the star-thronged galactic disk. These nuclear autographs had great theoretical significance, as they bolstered a revolutionary hypothesis advanced two decades earlier by four researchers working at the California Institute of Technology. In 1957, Britishers Fred Hoyle, Margaret Burbidge, and Geoffrey Burbidge, and American William Fowler had asserted that all elements heavier than helium are forged in nuclear reactions in the interiors of stars. Now that theory had a rare piece of supporting evidence. Because the aluminum isotope would decay to invisibility within a few million years, the substance Jacobson's team detected had to have come into being fairly recently, in cosmic terms. As the Caltech group had predicted, the elements probably were spewed into space by novae or by supernova explosions. Eventually they would be incorporated into new stars.

A spectacular supernova that blossomed in the Large Magellanic Cloud in 1987 corroborated Jacobson's observation. The aging Solar Maximum Mission satellite and assorted high-altitude balloons, keeping a vigil over supernova 1987A, recorded among its output gamma rays bearing the distinctive spectral lines of cobalt isotopes decaying into iron atoms. Since the half-life of cobalt is a matter of months, the discovery was a further validation of the theory of nucleosynthesis.

UNSOLVED MYSTERIES

At least two key questions in gamma ray astronomy remained unanswered. One was the cause of the enigmatic radiation bursts. After Klebesadel's report, scientists at Goddard and elsewhere had succeeded in gaining confirmatory measurements and spectral information that led them to believe the bursts were generated by neutron stars. Additional data from the Japanese satellite *Ginga,* launched in 1987, allowed astrophysicist Don Lamb and colleagues at the University of Chicago to carry out detailed computer analyses of the bursts. In 1988, they reported that the radiation must issue from regions dominated by magnetic fields more than a trillion times stronger than Earth's. Almost certainly, Lamb concluded, the dynamo at the heart of such a field would be a neutron star. However, he was at a loss to say how, exactly, the field could unleash such torrents of gamma rays. Astronomers hope that NASA's Gamma Ray Observatory, a seventeen-ton satellite set to fly on the space shuttle in 1990, will clarify the picture. The four instruments that have been designed for the orbiter will collect a wide range of gamma radiation and pin the location of sources to within a fraction of a degree, in addition to continuously watching for bursts.

The second pressing issue is the meaning of the background gamma chatter. Since this diffuse gamma radiation flows from all directions, astronomers assume that it is associated with the cosmic rays that also bombard the Earth

KEEN SENSORS OF THE COSMIC BOMBARDMENT

Through the years of the twentieth century, many different types of observing instruments covering the entire electromagnetic spectrum have joined the venerable optical telescope as surveyors of the cosmic spectacle. But the most fitting emblems of the new astronomy may be an odd assortment of devices, shown here and on the following pages, that look beyond electromagnetic radiation altogether. Their common feature is a focus on events and phenomena of the highest energies yet encountered.

Among the chief objects of study in high-energy astronomy are cosmic rays, which are not rays at all but streams of charged particles hurtling through space at tremendous speeds. They consist almost exclusively of atomic nuclei, primarily of hydrogen but of many other elements as well. Analyses of element distribution and of the particles' energy levels give astrophysicists clues to their origins—most likely supernovae, pulsars, and binary star systems, each capable of propelling matter with prodigious force.

The first cosmic ray particles detected were actually so-called secondary particles—fragments resulting from the collision of primary particles with molecules in Earth's atmosphere. To understand more about primary particles, instruments had to be sent beyond the atmosphere. In 1985, the shuttle *Challenger* carried in its cargo bay one of the most sophisticated such detectors, the Cosmic Ray Nuclei Experiment *(below)*. Its sensors employed several strategies to ensure that only primary particles were measured. Accurate timing of detections at identical scintillation counters, for example, established whether a particle was moving upward or downward, and readings across a central radiation detector indicated its trajectory. In this manner, fragments of particles that might have passed through the shuttle or the atmosphere could be ignored.

Although their trajectories appear straight within the detector, particles are easily deflected by magnetic fields such as Earth's—lower-energy particles *(wavy line, right)* more so than higher ones *(straighter line)*. Thus, precise cosmic ray sources remain a mystery.

THREE DETECTORS IN ONE

The cross-sectional diagram at right details the passage of a cosmic ray particle through the three types of detectors that make up the Cosmic Ray Nuclei Experiment. Penetrating the top dome, the particle enters the first of two Cerenkov counters *(1)*, each of which contains a gaseous mixture of nitrogen and carbon dioxide. The particle temporarily excites gas molecules along its path, causing a release of energy in the form of photons of light *(2)*. The photons ricochet off the reflective inner walls until they register at photomultiplier tubes *(3)*, which convert the light into a more easily analyzed electric signal whose intensity is directly proportional to the particle's energy level. A scintillation counter *(4)* then determines the particle's elemental identity through similar measuring of emitted light. As it traverses the central section, known as a transition radiation detector *(5)*, the particle emits x-rays that ionize gas molecules inside a series of narrow bands. Registering the resulting flashes, wires within these bands not only gauge the particle's energy but also trace its trajectory by pinpointing its position at each succeeding band. Duplicates of the scintillation and Cerenkov counters help confirm readings and rule out undesirable signals.

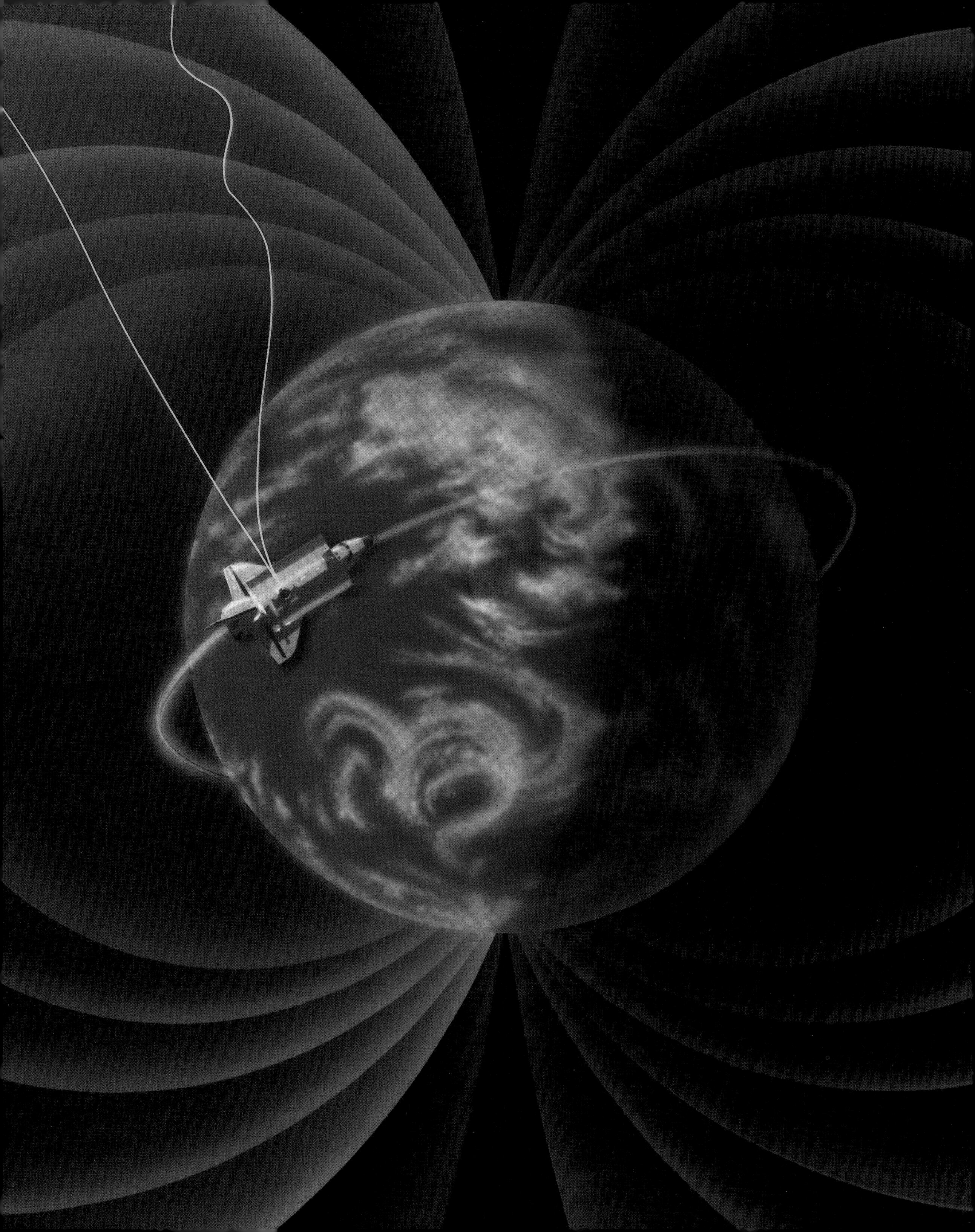

nucleus, two of them convert into neutrons by giving off positrons and neutrinos.

Supernova neutrinos have a different history. As a star's core collapses, the tremendous gravity forces electrons to merge with protons, again forming neutrons and generating neutrinos, in this case at extremely high energies. Because nothing impedes them, the neutrinos are the first particles to be ejected in the subsequent supernova explosion, flying outward ahead of the shock wave at or near the speed of light, even before light itself can penetrate the supernova's surrounding cloud of matter.

Instruments to detect these mysterious particles must be buried at least 2,000 feet deep, so that all other forms of radiation from space, even gamma rays, will be blocked. And because neutrinos interact only rarely with other matter, many billions of them must pass through the detector for a handful to be recorded. Despite the odds, two devices housed in mines—one near Kamioka, Japan, the other beneath the shores of Lake Erie—proved the existence of supernova neutrinos in 1987 when they captured a high-energy burst from supernova 1987A, the stellar cataclysm that occurred 170,000 years ago in the neighboring galaxy known as the Large Magellanic Cloud, almost a day before its brilliant flash reached Earth.

A Tank for Tracking Neutrinos

The neutrino tank diagramed at right is a highly simplified version of the detectors in Japan and America *(below)* that recorded the shower of neutrinos that passed through the planet from supernova 1987A. Each tank is filled with purified water and its walls lined with photomultiplier tubes. On rare occasions during a neutrino shower, an incoming neutrino *(1)* crashes into a proton *(2)* in the water, knocking free an electron *(3)*. The electron briefly speeds forward in the same direction as the neutrino, emitting a ring of visible radiation centered on its flight path. Although the electron almost immediately comes to a stop, some of the photons *(4)* continue on to be registered by the photomultiplier tubes *(5)*. The position of the ring of photon hits enables physicists to determine the neutrino's trajectory, which points straight back to its source—here, the Large Magellanic Cloud, visible only from the Southern Hemisphere. Neutrino tanks also detect antineutrinos, which knock loose positrons. But because positrons scatter in random directions, the light they emit provides no tracking clues.

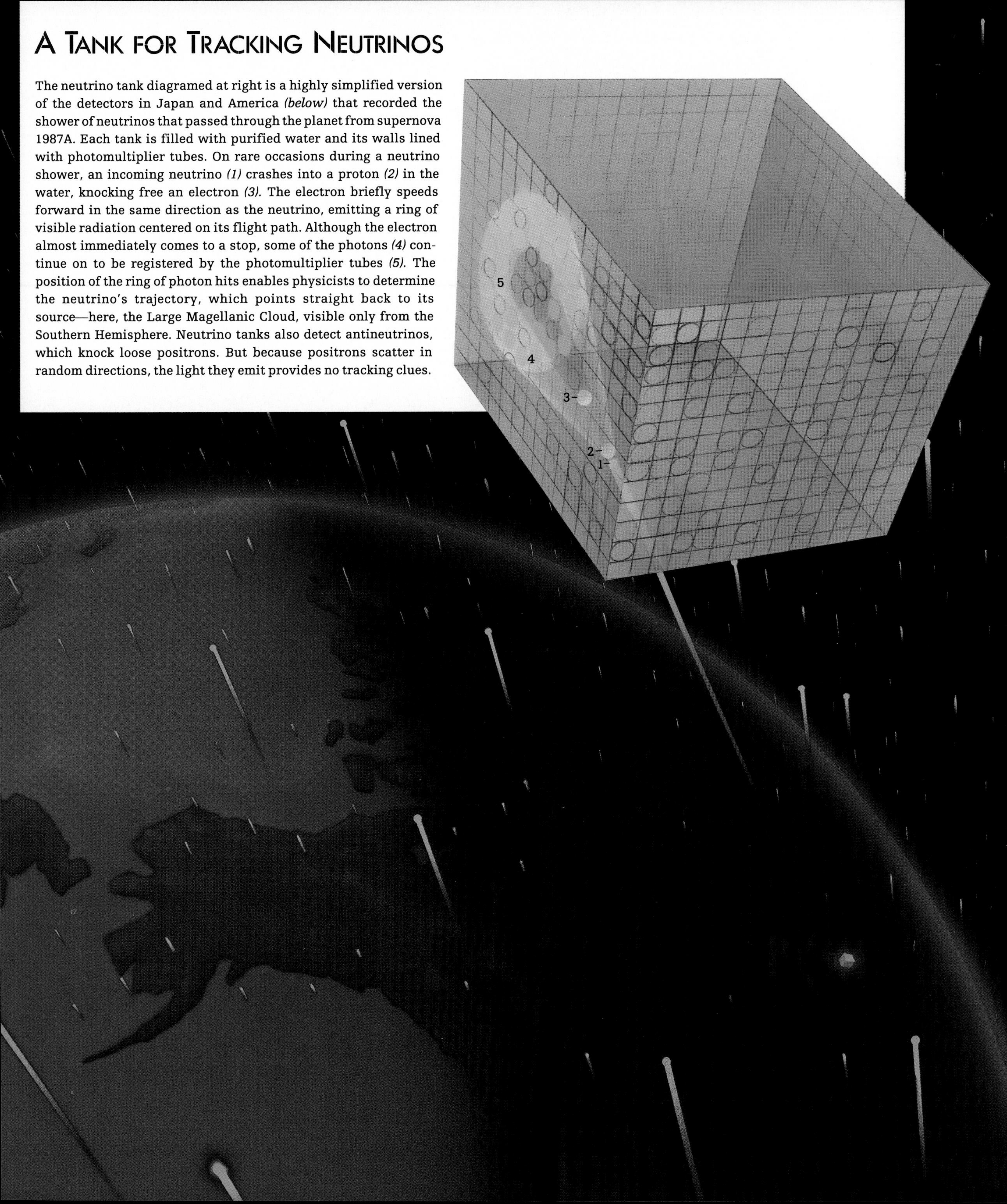

Ripples in the Fabric of Space

Among the many cosmological oddities predicted by Einstein is a form of radiation even more inscrutable than the neutrino: gravity waves. The theory of general relativity conceives of space and time as a continuum that can best be imagined as a firm rubber sheet, curved and warped by the gravitational force of celestial bodies. Any acceleration or jostling of very massive objects—the vibrations of a black hole, say, or a supernova explosion—should, theoretically, generate ripples in that sheet. These waves of gravitational energy, traveling at light speed, would temporarily compress space in one direction and stretch it in a perpendicular direction, but because the fabric of space is so stiff, the distortions would be nearly imperceptible by the time they reached Earth; Einstein himself thought they would never be spotted.

Modern astrophysicists are more optimistic. Their goal is to devise instruments that can detect changes in the distance between objects—on the order of a thousandth the diameter of a proton—caused by passing gravity waves; the greater the distance measured, the better the chance of capturing any distortion. The most promising design *(right)* would make measurements with laser beams bounced back and forth between mirrors nearly two and a half miles apart.

To improve their chances further, U.S. scientists plan to link two such detectors electronically, one in the east and one in the west: Local signal noise, different in each place, could thus be identified and removed, leaving a clean slate for gravity waves.

To Catch a Wave

The detector diagramed schematically below employs laser beams to spot the distance-distorting effects of gravity wav[es]. Each of its two perpendicular arms—corresponding to the t[wo] directions in which gravity waves alter the fabric of space—consists of a pair of heavy mirrors *(1)* at either end of a vacuum pipe *(2)* two and a half miles long; the mirrors and pip[e] are suspended to help isolate them from local vibrations. A beam-splitting mirror *(3)* set at the junction of the two arms divides light from a laser *(4)* into two beams, directing one down each pipe. The mirrors are specially constructed to allow the beams to resonate back and forth many times—maximize the distance traveled—before passing back t[o] the beam splitter, where they are reincorporated an[d] forwarded to a photodetector *(5)*. According to the principles of interferometry, the slightest chang[e] in the distance between mirror pairs will cau[se] a telltale alteration in the signal recorded [by] the photodetector *(right)*.

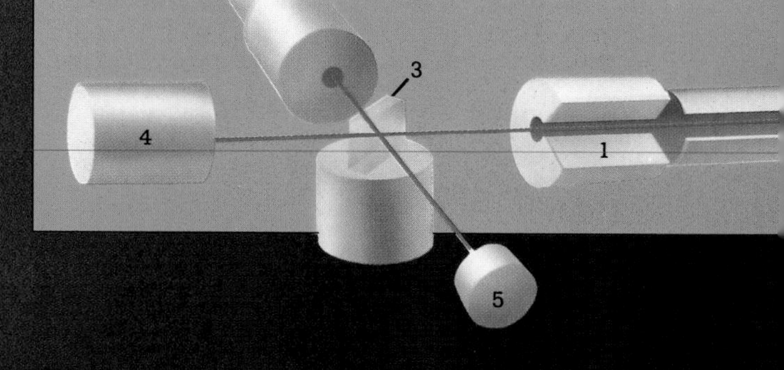

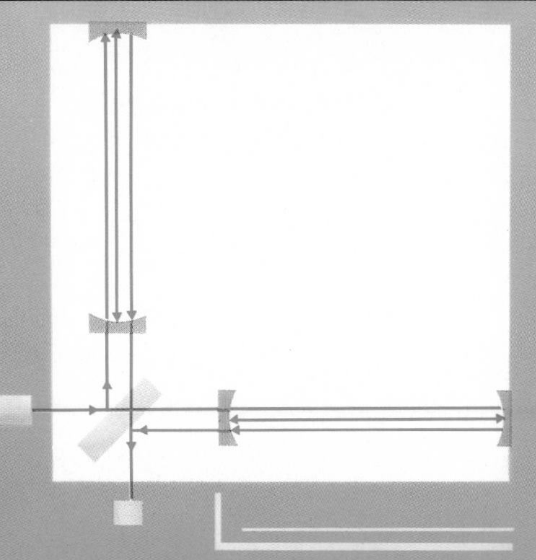

A quiet detector. At rest, the length of each detector arm is steady. Thus, when the two halves of the split laser beam rejoin after traveling back and forth down each arm *(red lines),* the wave peaks of one align with the troughs of the other and cancel each other out: The detector records no signal *(graph).*

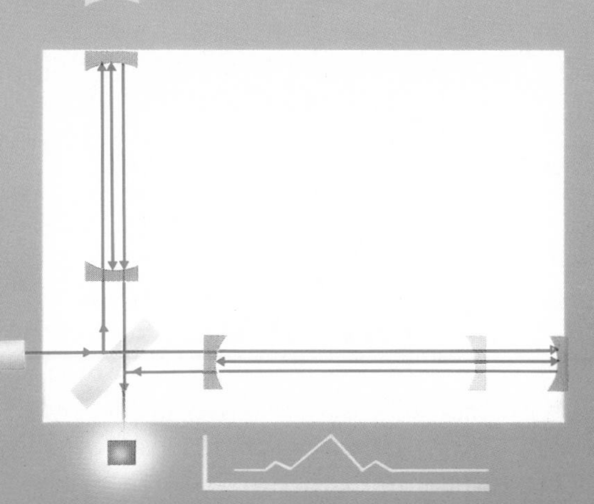

Wave interference. A passing gravity wave slightly shortens one arm and lengthens the other, nudging each outer mirror in a different direction. As a result, the peaks and troughs of the reunited beam halves, instead of aligning and canceling, now interfere with each other in a way that reinforces the beam, generating a signal picked up by the photodetector.

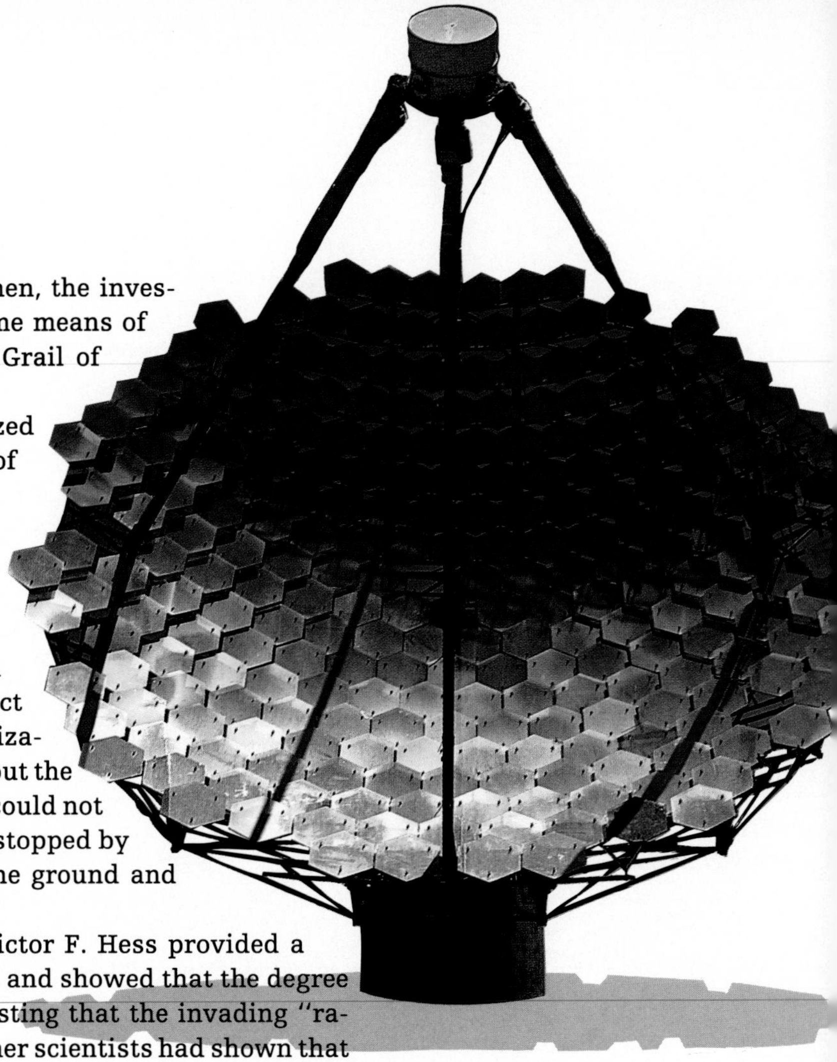

from all parts of the sky. To some scientists, then, the investigation of gamma radiation represents the prime means of tracking down what has been called the Holy Grail of astrophysics: the origin of cosmic rays.

The existence of cosmic rays has been recognized only in the last hundred years. Around the turn of the century, a British physicist named Charles T. R. Wilson set up an experiment to show that radiation from outer space was responsible for ionizing Earth's atmosphere—that is, the radiation was knocking electrons off neutral atoms. Installing his equipment in a Scottish railroad tunnel and working at night to minimize contact with trains, Wilson measured the degree of ionization in the tunnel's air. To his surprise, it was about the same as that outside. This meant that radiation could not be the culprit, since even gamma rays would be stopped by the rock of the tunnel. What could penetrate the ground and remain so energetic?

In 1912, Austrian-born American physicist Victor F. Hess provided a clue when he took instruments aloft in a balloon and showed that the degree of ionization increased at high altitudes, suggesting that the invading "radiation" came from outer space. By the 1930s, other scientists had shown that the mysterious invaders were influenced by Earth's magnetic field and were probably particles, not rays.

Eventually, physicists learned that the particles usually carry a positive charge; moreover, the interlopers seem to travel at almost the speed of light and have tremendous kinetic energies of greater than a trillion electron volts. Crossing the vastness of interstellar space, the rays meander away from a straight course, tugged hither and yon by the low-level magnetic field that permeates the universe. But when they arrive at the Earth, they still retain the energy associated with their enormous speed, plunging through thousands of feet of water or rock on the planet's surface.

Despite this enormous penetrating power, few "primary" cosmic rays actually reach the ground. Instead, they collide with atoms in the atmosphere and give rise to showers of "secondary" particles, which in turn initiate cascades of electrons and muons, subatomic particles that help to hold atomic nuclei together. These showers give off a faint blue light called Cerenkov radiation. On dark, clear nights, ground-based arrays of mirrors coupled with photomultiplier tubes can actually detect this glow *(above).*

For a time during the 1930s and 1940s, cosmic ray research was the domain of physicists rather than astronomers. But in the 1950s, when a far more powerful tool for studying subatomic events was created, namely the high-energy particle accelerator, physicists abandoned atmospheric studies and astronomers gradually took over. Deploying experiments by balloon and

Astronomers prospect for traces of high-energy gamma rays with sensitive optical telescopes like the ten-meter reflector at Arizona's Whipple Observatory. Faint flashes of blue Cerenkov light, which are produced by the gamma rays' interaction with atoms in Earth's atmosphere, are reflected by the telescope's 248 paraboloidal mirrors onto central detectors that convert the light into electrical pulses.

space probes throughout the 1960s and 1970s, researchers began to draw a tentative portrait of cosmic radiation.

About 98 percent of cosmic rays, astronomers have found, are in fact atomic nuclei stripped of their electrons, representing all elements from hydrogen to uranium. By far the most common cosmic rays to reach Earth are those with energies between one billion and one trillion electron volts, which strike at the rate of about 1,000 per two to three square feet per second. Going up the scale of energy, the rays become more and more scarce. Each year, fewer than a dozen rays at 10 million trillion eV, the most energetic yet detected, hit a square mile of the planet.

Hoping to find out what creates this steady rain of particles, researchers in 1979 loaded the NASA satellite *HEAO-3* with instruments for determining the charge, energy, and trajectory of high-energy cosmic rays. *HEAO-3* spectral measurements picked up cosmic nuclei for various elements near Earth, including hydrogen, helium, carbon, oxygen, neon, magnesium, silicon, calcium, and iron; their relative abundance matched the abundance of elements in the Solar System and was not dependent on their energies, which fell into a narrow band between 1 and 25 billion electron volts, or gigaelectron volts (GeV). This seemed to indicate that the particles were born in the same evolutionary processes that created other matter, leaving the issue of their extreme speeds a mystery. Six years later, a Spacelab II experiment on the *Challenger* shuttle clouded the picture by determining that the relative abundance of some elements varied at higher energies, raising questions about both the sources of the particles and the mechanisms that accelerated them.

The best explanation astronomers have come up with so far is that the same cataclysmic or highly energetic processes that produce gamma rays also produce cosmic rays. Supernovae are likely sources, as are pulsars and binary x-ray stars. According to the supernova scenario, shock waves from an exploding star may smack into charged particles that float through space as part of normal interstellar matter. Pumping energy into the particles, the waves fling them across space. Pulsars might hurl particles away from their rapidly spinning magnetic fields, and x-ray binaries might slough off particles as part of the accretion process. Exciting evidence in support of the last hypothesis has come from Cerenkov detectors that have registered secondary gamma radiation apparently linked to cosmic particles from the binary systems Cygnus X-3 and Hercules X-1, among others. The hunt for the cosmic grail has just begun, and astronomers are pressing on, perfecting ever more elaborate sensing devices attuned to higher and higher energies.

IN SEARCH OF GHOSTLY PARTICLES

Cosmic ray researchers are concentrating on the story of the cosmos as told by flying particles, but beyond that, there are at least two other versions of the sky: the universe according to neutrinos, and the universe according to gravity waves.

Neutrinos were born as a theoretical concept in 1930, when the eminent physicist Wolfgang Pauli hypothesized that atoms would emit these infinitesimally small, chargeless elementary particles as they decayed. In the 1950s, the first neutrino was detected—no small feat, since these particles hardly interact with matter at all. Neutrinos may or may not have mass, an issue that preoccupies cosmologists, since the mass, if it exists, supplied by vast numbers of neutrinos would be sufficient to one day pull the universe back in upon itself. Unfathomably numerous, the tiny particles swim through the universe at the speed of light, able to dart across trillions of miles of space without brushing another particle. Scientists believe that some were created a split second after the inception of the universe along with a host of other matter and antimatter particles. Other neutrinos may have come into being in the nuclear fires of stars and were emptied into space in supernova explosions.

Because neutrinos travel the entire universe, they likely have fascinating tales to tell. However, their very ease of movement poses serious impediments to would-be listeners, since they pass as effortlessly through Earth itself as through empty space. Contemplating this dilemma in the 1960s, a University of Pennsylvania astronomer named Ray Davis dreamed up a strange device for capturing the elusive particles. The receptacle he chose was a huge tank hewn out of rock almost a mile underground in the abandoned Homestake mine in Lead, South Dakota. There, the constant noise of cosmic rays is blocked, and the whisper of neutrinos can be heard. Some 100,000 gallons of tetrachloroethylene, a dry-cleaning fluid composed largely of chlorine atoms, fills the tank. On rare occasions, a neutrino smashes into one of these chlorine atoms; the collision produces a detectable atom of radioactive argon gas.

Davis aimed to collect solar neutrinos, which by all rights should be numerous. However, by 1981 it was clear that the apparatus was routinely picking up only about a third the number of neutrinos that theorists predicted it should. Moreover, even the neutrinos that did leave their mark revealed little, since there was no way to divine their flight path.

Serendipitously, two similar tanks built for an entirely different purpose

Sixty-seven large mirrors, each housed in a protective metal drum seven feet long and eight feet wide, form the Fly's Eye Cosmic Ray Observatory at Little Granite Mountain, Utah. Mimicking the multiple facets of an insect's eye, the mirrors angle toward different regions of the night sky to collect Cerenkov radiation triggered when high-energy particles known as cosmic rays collide with atoms in the upper atmosphere. Observers use the data to compute the energy of the cosmic rays and the directions they come from.

provided the first real glimpses of astronomically significant neutrino activity. Completed in 1986, the tanks were supposed to test the notion that protons do not persist forever but instead decay after about 10^{19} years. Proton decay bore heavily on the so-called grand unification theory of physics, which attempts to reconcile all the primary forces of nature except gravity. Experimenters reasoned that if they observed a sufficient number of protons they might eventually see the death of one, which would be accompanied by a flash of Cerenkov radiation. Once again, the investigators placed their equipment deep underground, where it would be blanketed against bothersome cosmic rays. In Kamioka, Japan, about 300 miles west of Tokyo, thousands of gallons of purified water filled a tank whose sides were lined with photomultiplier tubes. In Fairport Harbor, Ohio, on the shores of Lake Erie, a similar array was set up in a former salt mine. Over the course of several years, the researchers have monitored the tanks for episodes of proton decay. As yet, they have caught no protons in the act, but their efforts have not gone for naught. Even as the tanks were being constructed, investigators realized their equipment might also register passing neutrinos if the particles grazed a proton and sent it flying.

SWIFT, SILENT HERALDS

This was precisely what happened in the hours preceding the spectacular appearance of supernova 1987A. Almost a full day before the supernova became visible, the Kamioka tank racked up the passage of eleven neutrinos within 12.5 seconds. Apparently simultaneously, eight neutrinos sped through the Lake Erie detector in about six seconds. Since the tanks pick up only a tiny fraction of the torrent of neutrinos released toward Earth, the numbers spotted and the energies involved implied an explosion of unimaginable power—one that released a hundred times more energy than the Sun will produce in its entire ten-billion-year lifetime. Mystified scientists did not recognize the neutrino sightings as being connected with the supernova until some days later, after observers spotted the explosion with optical telescopes. But subsequent examination of the data has suggested to astronomers that the neutrinos began their cosmic journey during the collapse of the blue supergiant star that was the supernova's precursor.

Some astronomers dream of a worldwide network of underground tanks and a full-fledged program to study neutrinos of different energies. Indeed, several detectors are currently in the planning stages, including one to be built in the waters of the Pacific Ocean off the coast of the island of Hawaii—a kind of floating cage of photomultiplier tubes. These may turn neutrino astronomy into a genuine observational science rather than a purely theoretical pursuit. In addition to providing evidence about the eventual fate of the universe, neutrino detectors should allow astronomers to detect supernovae that otherwise might never be noticed because their radiation at visible wavelengths is blocked by interstellar dust.

As difficult as it is to capture spectral entities such as neutrinos, the task looks simple next to that of detecting the most elusive phenomenon of all: gravitational waves. Predicted by Einstein on the basis of his general relativity theory, gravitational waves are supposed to be stirred up by massive objects as they whirl, vibrate, collapse, or explode. In effect, such objects jiggle the jelly of space-time, sending ripples outward at the speed of light. The waves, which can be neither slowed nor stopped by matter, have no effect along their primary direction of movement, but they do vibrate both sideways and up and down. However, since the strength of gravitational waves reaching Earth from even a gargantuan event is fantastically small, the waves are almost impossible to measure: The wave generated by a nearby supernova, for instance, would perturb matter on the planet by about one-thousandth the diameter of a proton.

The scientist who finds gravity waves will advance the cause of physics and astronomy by a giant step. For one thing, the discovery would greatly support Einstein's relativistic portrait of the universe. For another, the kind of energy carried by the waves would provide an entirely new view of the sky, unhampered by intervening matter or radiation. Einstein himself doubted that anyone would ever find gravity waves, but a few scientists dare to think otherwise. Contemplating the subject in the late 1950s, Joseph Weber, a University of Maryland physicist, became convinced that he could capture the waves with an apparatus he called a gravity wave antenna. A solid cylinder of aluminum, the detector would be compressed slightly by a passing gravitational wave. This accordion-like movement would disturb attached piezoelectric crystals, which generate an electric current when under mechanical stress.

The first antenna Weber assembled was, in a way, too sensitive; it could be fooled by passing traffic, earth tremors, lightning, and so on, and had to be insulated against such distractions. But by 1969, Weber was certain that he had recorded on two antennas gravity waves from the center of the galaxy. Astronomers around the United States hurriedly constructed their own antennas, eager to partake of the historic discovery, but their excitement quickly faded. When they failed to get similar results, they began questioning whether Weber had not been deluded.

The fault was less Weber's than the technology's. Currently, even the most advanced detectors are too dull to sense the subtle waves. Still, astronomers persist in their efforts, since by the equations of general relativity a host of massive objects in the universe should be sending out gravity radiation. In the late 1980s, several groups of researchers were making headway in improving instrumentation. At Stanford, a team headed by William Fairbank designed a supercooled antenna, a five-ton aluminum bar bathed in liquid helium to keep it a shade above absolute zero, where all movements of its atoms would cease. Suspended on special springs in a windowless building, the antenna is virtually interference-free and can record movements as slight as one 30-millionth the diameter of a hydrogen atom.

Other researchers have placed their hopes on another type of detector called a laser interferometer, in which two mirrors are suspended in a vacuum at a distance from one another and their positions constantly measured with a laser beam *(pages 84-85)*. With the passage of a gravity wave, the mirrors should move, slightly widening or narrowing the distance between them, thereby perturbing the laser beam. Researchers have built two prototype interferometers in Europe and two in the United States, at MIT and Caltech. With a 130-foot spread between mirrors, the device at Caltech is the world's largest. However, interferometers have a long way to go to surpass Stanford's antenna in sensitivity. In the next phase of the research, U.S. investigators plan to base interferometers with two-mile-long arms in the Mojave desert and in Maine, which should make them 1,000 times more sensitive than their predecessors.

At the Jet Propulsion Laboratory in Pasadena, California, physicist Frank Estabrook and astronomer John Armstrong have planned an experiment with an even broader scope. Using the Galileo spacecraft scheduled for launch in late 1989, the test will be able to detect a passing gravity wave by measuring its influence on the microwave signal that tracks the probe from Earth. As part of the mission's normal procedures, scientists at JPL monitor the motion of the spacecraft by sending a radio signal to the craft and noting the change in

A Test of Relativity

A binary pulsar in the constellation Aquila, 15,000 light-years from Earth, provides astronomers with a celestial laboratory for investigating a phenomenon predicted by Einstein's general theory of relativity. The radio pulsar, known as PSR 1913 + 16, and its nonemitting companion orbit a common center of mass *(black cross)*, completing each circuit in about eight hours. In more than a decade of observation, scientists have detected a slight but steady decrease in the orbital period, a sign that the orbits are shrinking *(blue ellipses)*, which, in turn, indicates that the system is losing energy.

Astronomers attribute the energy loss to the radiation of gravity waves *(green)*. According to the general theory of relativity, such waves are emitted by any accelerated mass. The amount of gravitational radiation calculated for the pulsar and its companion (which accelerate to about 250 miles per second at their closest approach) is within one percent of the observed energy loss—compelling evidence that Einstein was right.

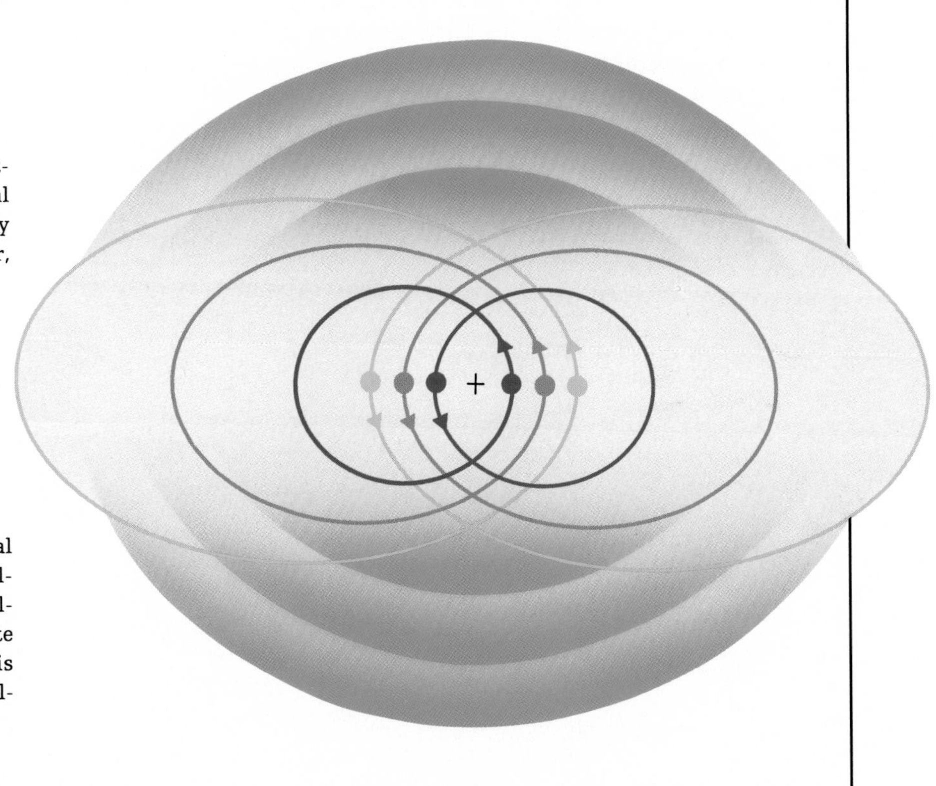

its return frequency; waves from an object moving away from the Earth should lengthen slightly, while those from an approaching object should shorten. In the gravity wave experiment, the Earth and the probe are widely separated test masses. If a gravity wave hits the Earth-spacecraft system, it should change the frequency of the transmissions as it shakes first one transmitter and then the other. The experimenters believe that this test should be sensitive enough to detect waves emitted by very massive objects, such as black holes.

Despite excruciatingly slow progress, gravity wave researchers are convinced that their hunt is not mere folly. For example, according to Joseph Taylor of Princeton and colleagues at the University of Massachusetts at Amherst, a pair of orbiting neutron stars in the constellation Aquila, one a pulsar, must be sending out gravity waves. Because a pulsar's signals are extremely regular and predictable, Taylor has been able to measure tiny differences in the star's orbit over time and has shown mathematically that the stars lag in their movement around each other precisely as they would if they were radiating away energy in the form of gravity waves *(page 91)*.

The race to look into unexplored regions of the cosmos is closely tied to technological breakthroughs on Earth, and those who create and master new devices will probably be the winners. For astronomers of all stripes, the goal has always been personal: to be the first to see what has never before been seen. But a second motive has also driven them to the frontiers—an innate desire to open new windows on the universe, a universe that, as Stanford's gravity wave detective Fairbank says, "always turns out to be more complicated than we originally think."

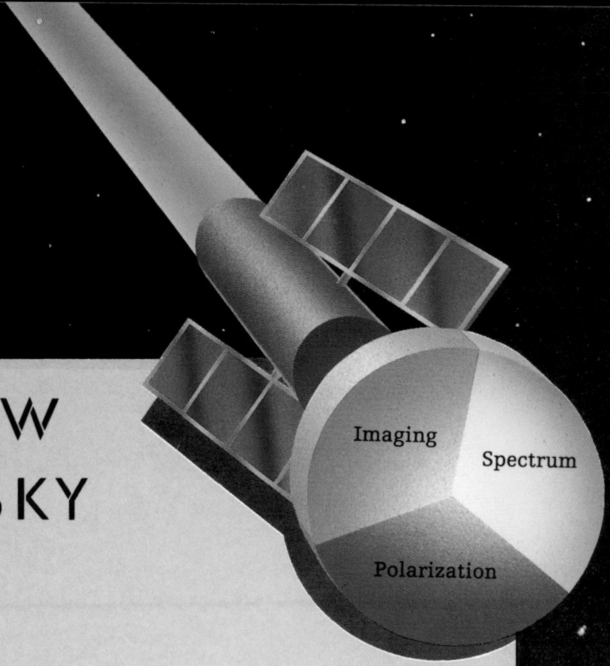

A HOST OF NEW EYES ON THE SKY

Peering into the heavens from the tops of remote mountains, leviathan eyepieces called telescopes—a name that comes from the Greek for "far seeing"—have long symbolized humanity's quest to know its place in the universe. That quest is now shared by telescopes of many kinds. Scrutiny of the cosmos, once limited to the narrow range of visible light, has come to embrace the entire electromagnetic spectrum and is approaching the very edge of time and space.

As indicated schematically above, the instrument attached to any modern telescope is concerned with one or another aspect of electromagnetic radiation: Instruments that analyze polarization, for example, give information about the orientation of the electric and magnetic energy of a wave. Other devices break up radiation into a spectrum of its constituent wavelengths. The detectors and instruments illustrated on the following pages are primarily used to record the spatial distribution of intensity, thereby creating images of otherwise invisible objects and revealing hidden aspects of objects already familiar in the optical realm. To this end, scientists have tailored their instruments to the type of radiation they wish to study. For example, capturing and focusing the faint, long-wavelength radio energy pulsing across the void from distant galaxies requires gargantuan reflectors. In contrast, very short wavelength gamma rays flung from black holes and neutron stars have such high energy that they cannot be focused at all. To photograph gamma ray sources, scientists must reconstruct the energy of each photon individually.

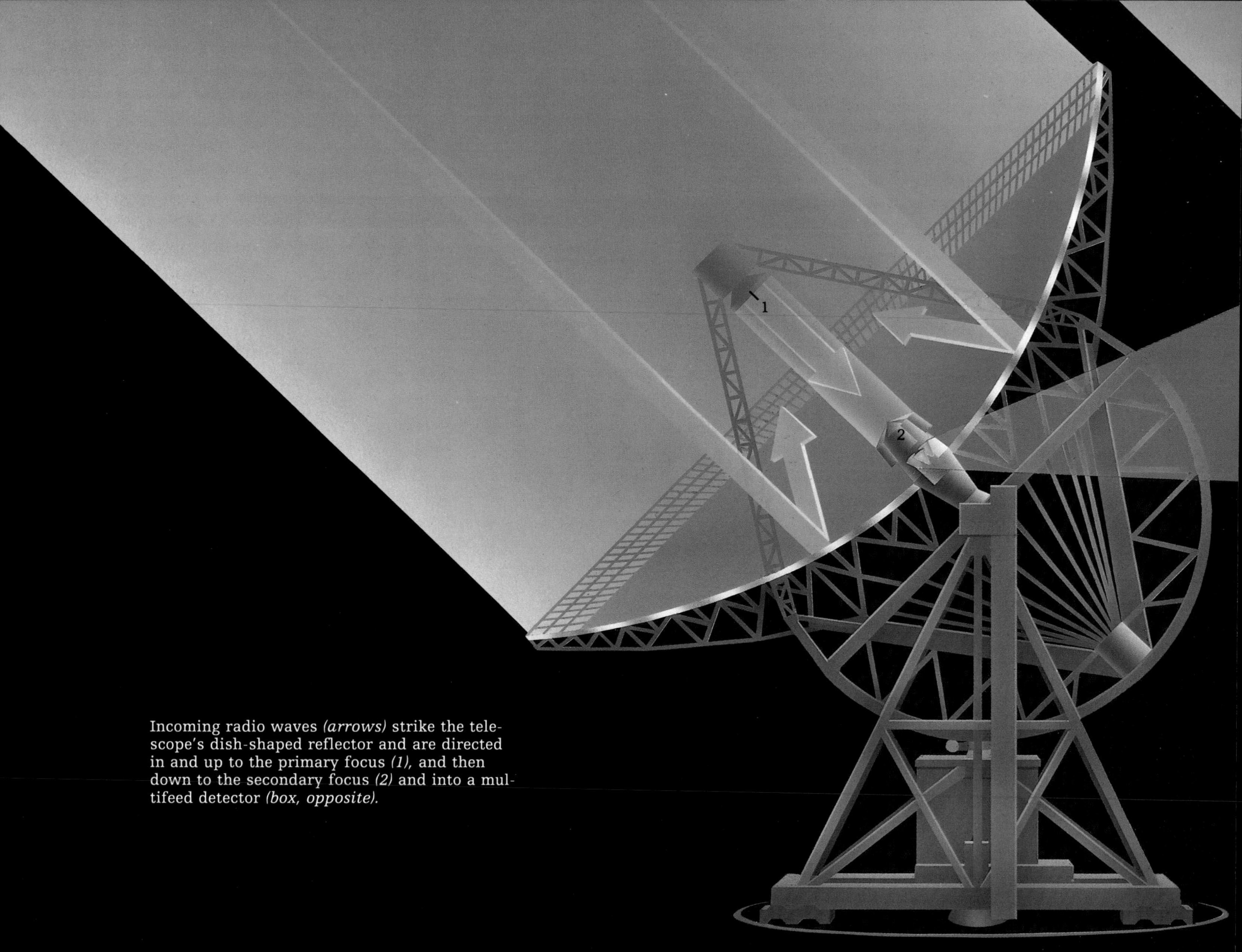

Incoming radio waves *(arrows)* strike the telescope's dish-shaped reflector and are directed in and up to the primary focus *(1),* and then down to the secondary focus *(2)* and into a multifeed detector *(box, opposite).*

SCANNING THE RADIO UNIVERSE

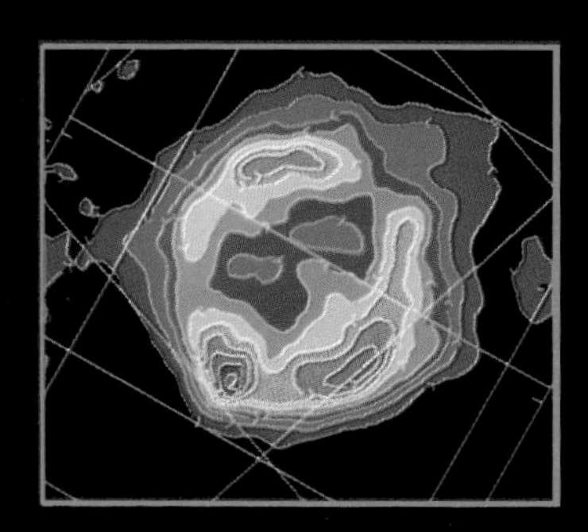

The larger a radio telescope's dish-shaped antenna, the better its ability to resolve detail in celestial sources. This gain in resolution is offset, however, by a certain loss of speed: A big telescope can measure radiation from only a small point on the sky, so to gather enough data to build up an adequate image of a larger region, the apparatus must be repositioned from one point to another.

One way around this dilemma is to attach devices that, in effect, allow the telescope to measure radiation from more than one point at a time. In Effelsberg, West Germany, the radio telescope shown in cross section above is equipped to do just that. At 328 feet in diameter and towering some twenty-seven stories high, it is the world's largest fully steerable telescope, capable of being rotated on two axes, to view virtually the entire sky in the Northern Hemisphere. The system described on the opposite page enables astronomers to exploit the instrument's capacities to produce radio images like the one of Cassiopeia A shown at left. Hot tatters of gas expanding from a supernova that exploded 300 years ago make Cassiopeia A the brightest radio source in the sky.

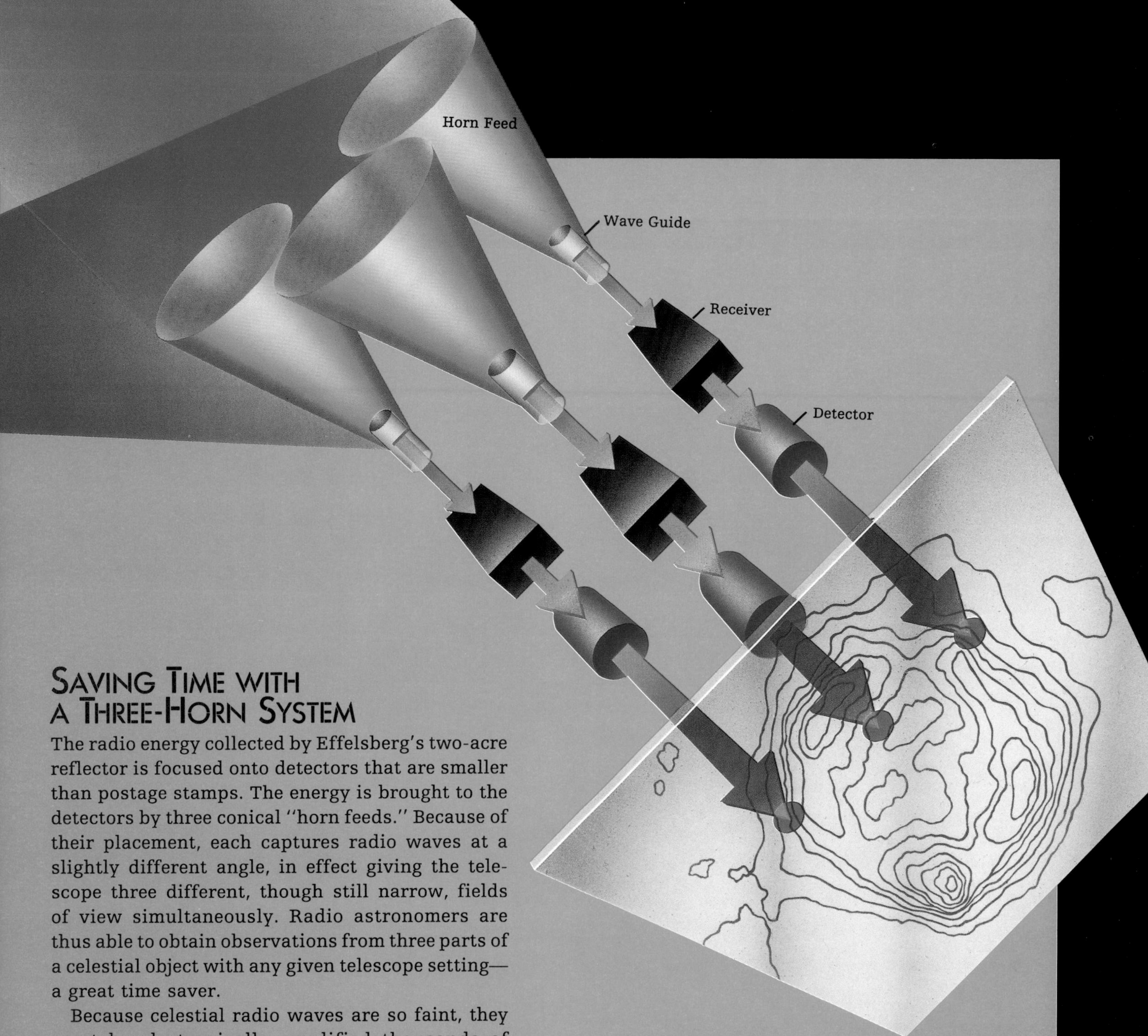

SAVING TIME WITH A THREE-HORN SYSTEM

The radio energy collected by Effelsberg's two-acre reflector is focused onto detectors that are smaller than postage stamps. The energy is brought to the detectors by three conical "horn feeds." Because of their placement, each captures radio waves at a slightly different angle, in effect giving the telescope three different, though still narrow, fields of view simultaneously. Radio astronomers are thus able to obtain observations from three parts of a celestial object with any given telescope setting—a great time saver.

Because celestial radio waves are so faint, they must be electronically amplified thousands of times. Each horn feed thus funnels radio signals into a waveguide *(orange cylinder),* where a central wire antenna, or probe *(red line),* converts the signal into an alternating electric current and passes it to the receiver *(brown box).* Amplified by the receiver, the signal goes to the detector *(purple cylinder)* and is converted into a direct current, which is then fed to a computer. When many points on the sky have been read in this way, the computer processes the combined signals into an image that can be displayed on a television monitor and photographed.

Infrared rays *(arrows)* enter the UKIRT and are reflected from the primary mirror *(1)* up to the secondary *(2)*, which bounces them back down through a hole in the center of the primary. Smaller mirrors direct the light to a box containing the array detector, whose workings are described on the opposite page.

Detecting the Impact of Photons

SPEED-READING AT INFRARED WAVELENGTHS

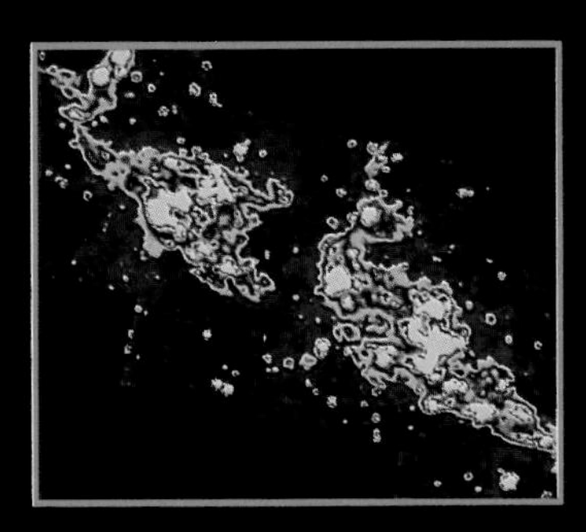

Peeking through clouds of gas and dust that would smother visible light, a smudge of infrared radiation in the picture at left heralds the birth of a new star inside the molecular cloud DR21. The image is the product of the United Kingdom Infrared Telescope (UKIRT), perched near the 13,796-foot summit of Mauna Kea on the island of Hawaii. With a 3.8-meter primary mirror, it is the world's largest instrument dedicated to infrared astronomy.

Until the early 1980s, infrared astronomers were handicapped by the fact that their telescopes were equipped with only one detector. To create a detailed picture of the sky, the telescope had to scan back and forth across an area of sky while the detector measured the infrared intensity at each point. New "array" detectors, however, can measure thousands of points at once. Using the old technology, the image of DR21 would have required several weeks to create; instead it took only a few hours.

The multilayered infrared array detector measures infrared intensity for more than 3,500 picture elements, or pixels, at a time; only nine are shown here. Incoming infrared photons *(blue arrows)* are absorbed by the detector's top layer of indium antimonide, a material sensitive to infrared energy. As the photons knock loose negatively charged electrons *(white balls)* from the indium antimonide, the electrons are attracted toward the second layer—indium antimonide that has been chemically altered, or "doped," to give it a positive charge. The electrons collect in a kind of electronic well between the middle and bottom layers, then flow through metal connections toward silicon transistors *(orange squares).* The transistors connect each well to a voltmeter, which measures the number of electrons collected during the exposure. Since each well corresponds to a pixel, a computer can translate these numbers into a final image that shows the level of infrared brightness at each point.

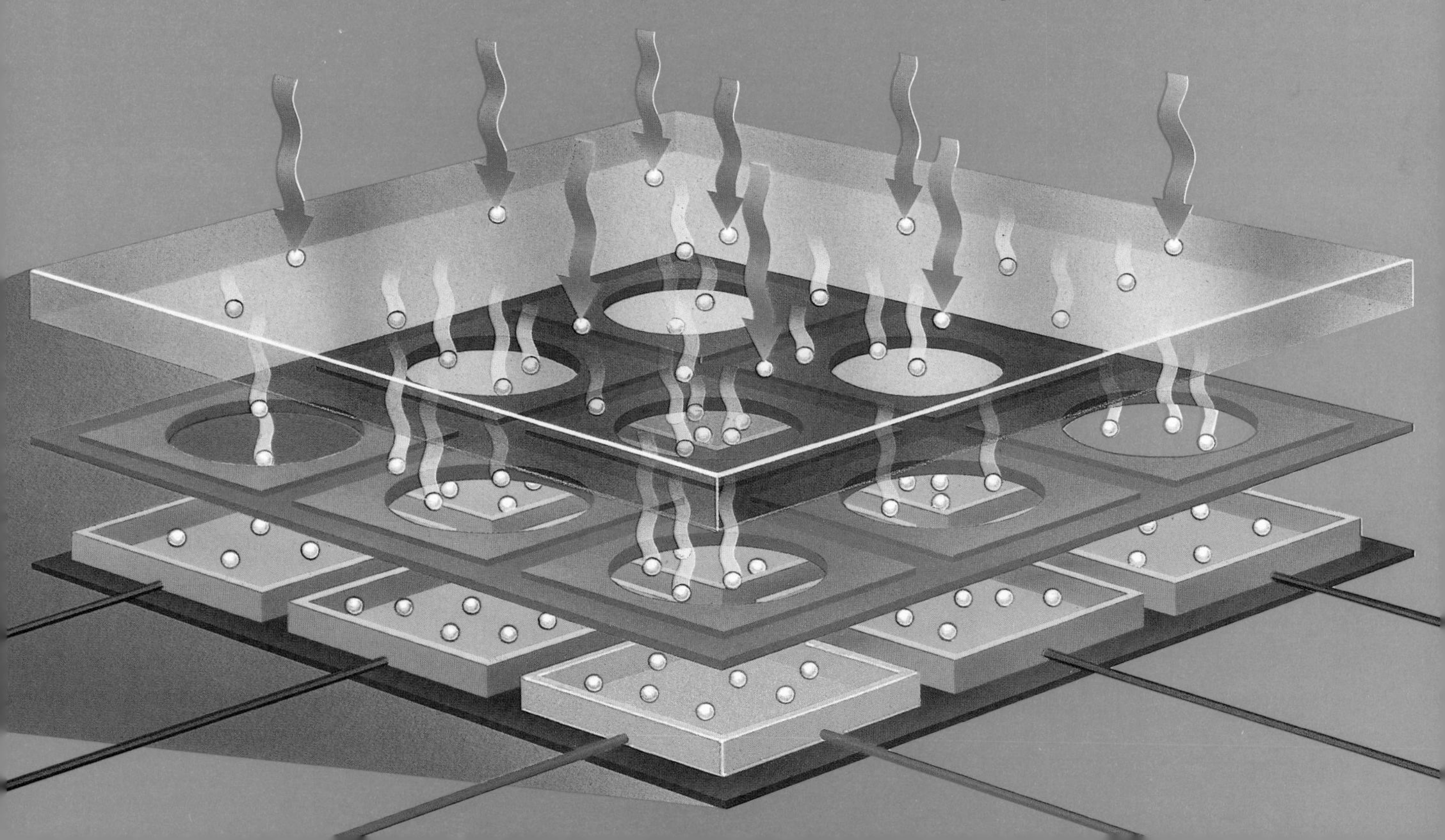

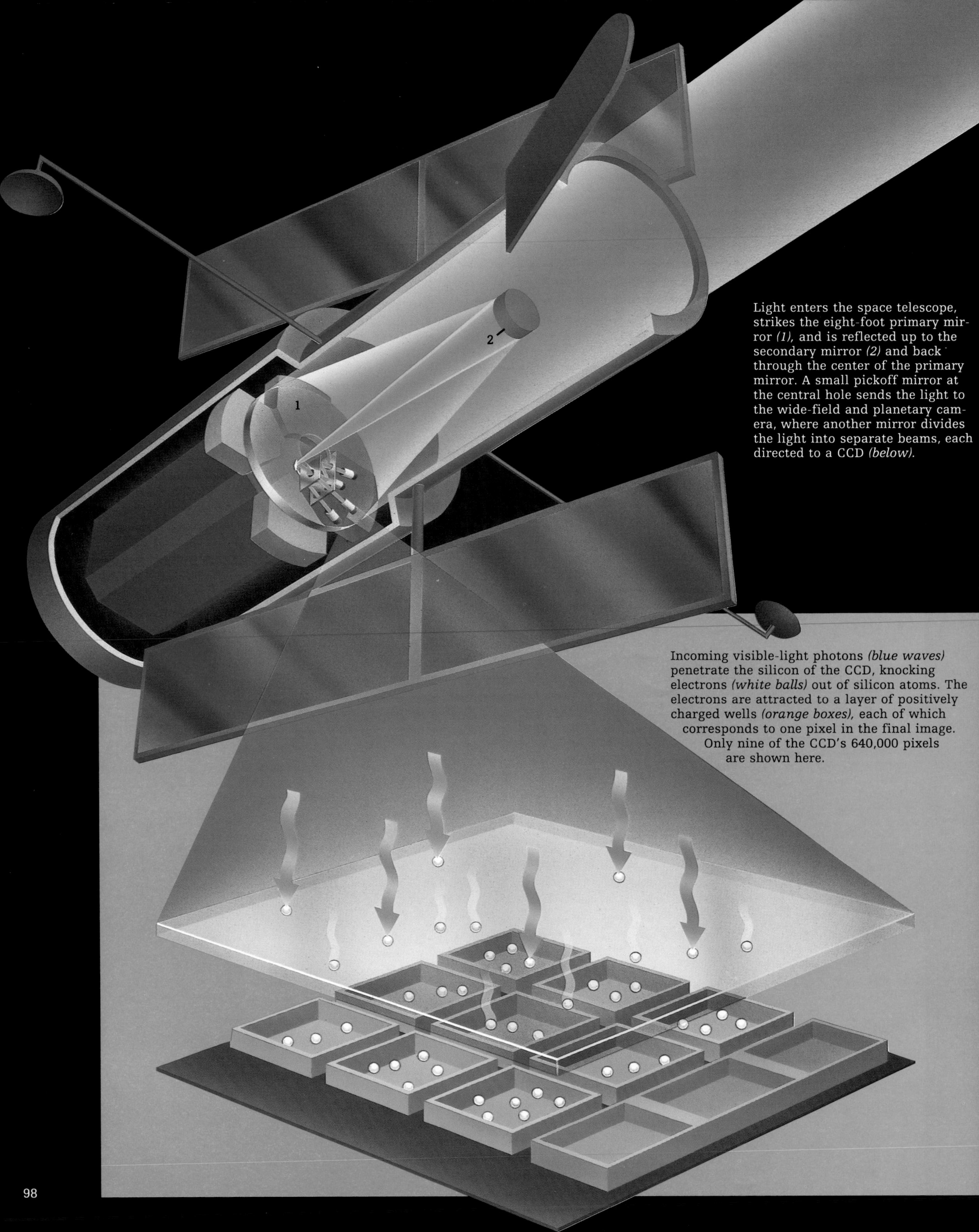

Light enters the space telescope, strikes the eight-foot primary mirror *(1),* and is reflected up to the secondary mirror *(2)* and back through the center of the primary mirror. A small pickoff mirror at the central hole sends the light to the wide-field and planetary camera, where another mirror divides the light into separate beams, each directed to a CCD *(below).*

Incoming visible-light photons *(blue waves)* penetrate the silicon of the CCD, knocking electrons *(white balls)* out of silicon atoms. The electrons are attracted to a layer of positively charged wells *(orange boxes),* each of which corresponds to one pixel in the final image. Only nine of the CCD's 640,000 pixels are shown here.

A Flawless Mirror on the Cosmos

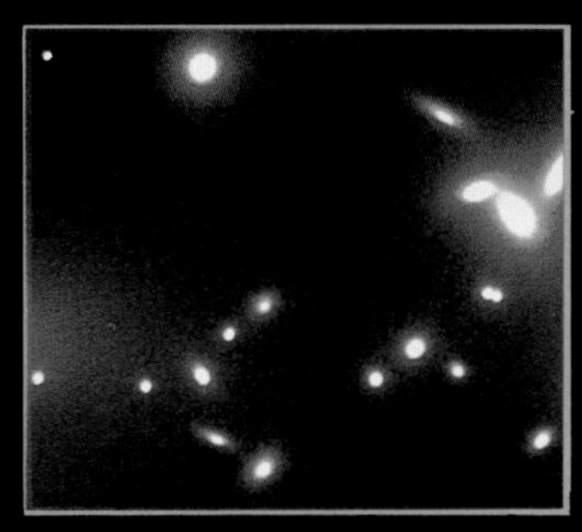

Looking at the sky through the light-distorting turbulence of Earth's atmosphere has often been likened to peering up from the bottom of a lake. The Hubble Space Telescope, due to be launched by NASA in 1990, will rise above that problem. Orbiting more than 350 miles above the atmosphere, the satellite will record visible-light images of unprecedented resolution. Such care has been taken in the manufacture of its eight-foot primary mirror that at the scale of the Unit ed States, the mirror's worst imperfections would b barely two inches high.

The Hubble will also carry a two-in-one instrumen known as a wide-field and planetary camera. Not in tended for full-sky surveys, the camera has a narrov field of view in either mode—less than one-tenth th diameter of the Moon for the wide-field mode, the dis of Jupiter for the planetary mode. The camera will ta the extraordinary capacities of detectors known a charge-coupled devices, or CCDs, silicon chips 10 times more sensitive than photographic film. A shown in the two pictures at left—simulations gen erated by computer for illustration—images fror the space telescope *(near left)* will reveal details in galaxy cluster that are only a blur in the ground-base version *(far left).*

Counting Out with a Charge-Coupled Device

The three steps below show in highly simplified form how a charge-coupled device captures photons and converts them into an electronic signal that can be transformed by computer into a detailed image. In principle, CCDs are similar to the infrared detectors that are described on pages 97-98: Each photon generates an electron, which is then trapped in a so-called well and counted. However, unlike the infrared detector, CCDs are made of silicon throughout. No layer of indium antimonide is needed for the generation of electrons because photons of visible light are much more energetic than infrared photons and can knock electrons directly out of the more tightly bonded silicon atoms.

To be counted, electrons must migrate, one row at a time, across the grid to the count-out row along the end. Positive voltages applied in succession to each row pull the negatively charged electrons from one row to the next, like buckets emptying in a brigade.

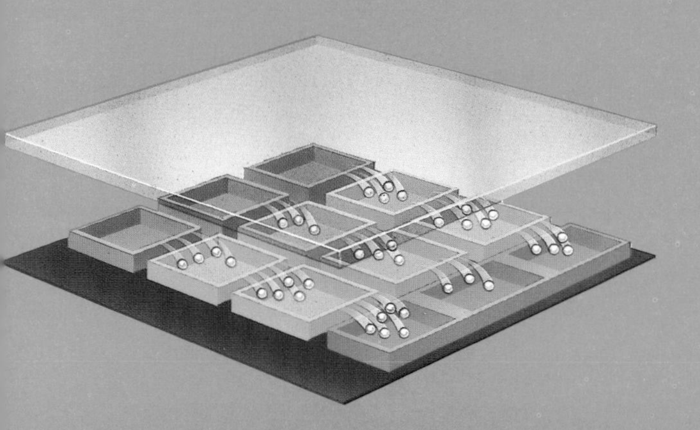

In the count-out row, electrons are pulsed out, one well—or pixel—at a time, and are counted by a special circuit that transmits the results to a computer for processing into an image. Here, one row of pixels empties as the remaining two rows wait their turn.

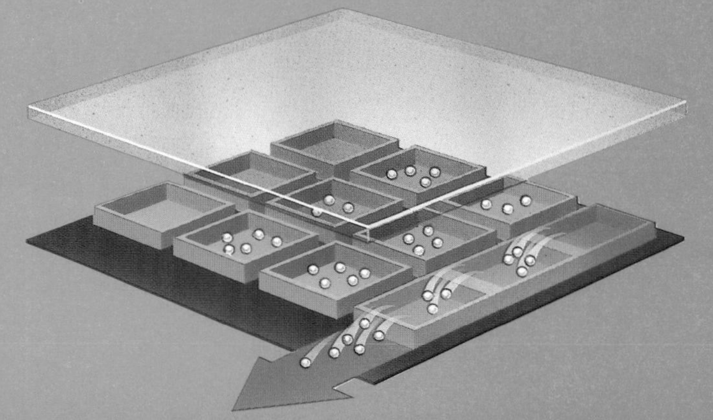

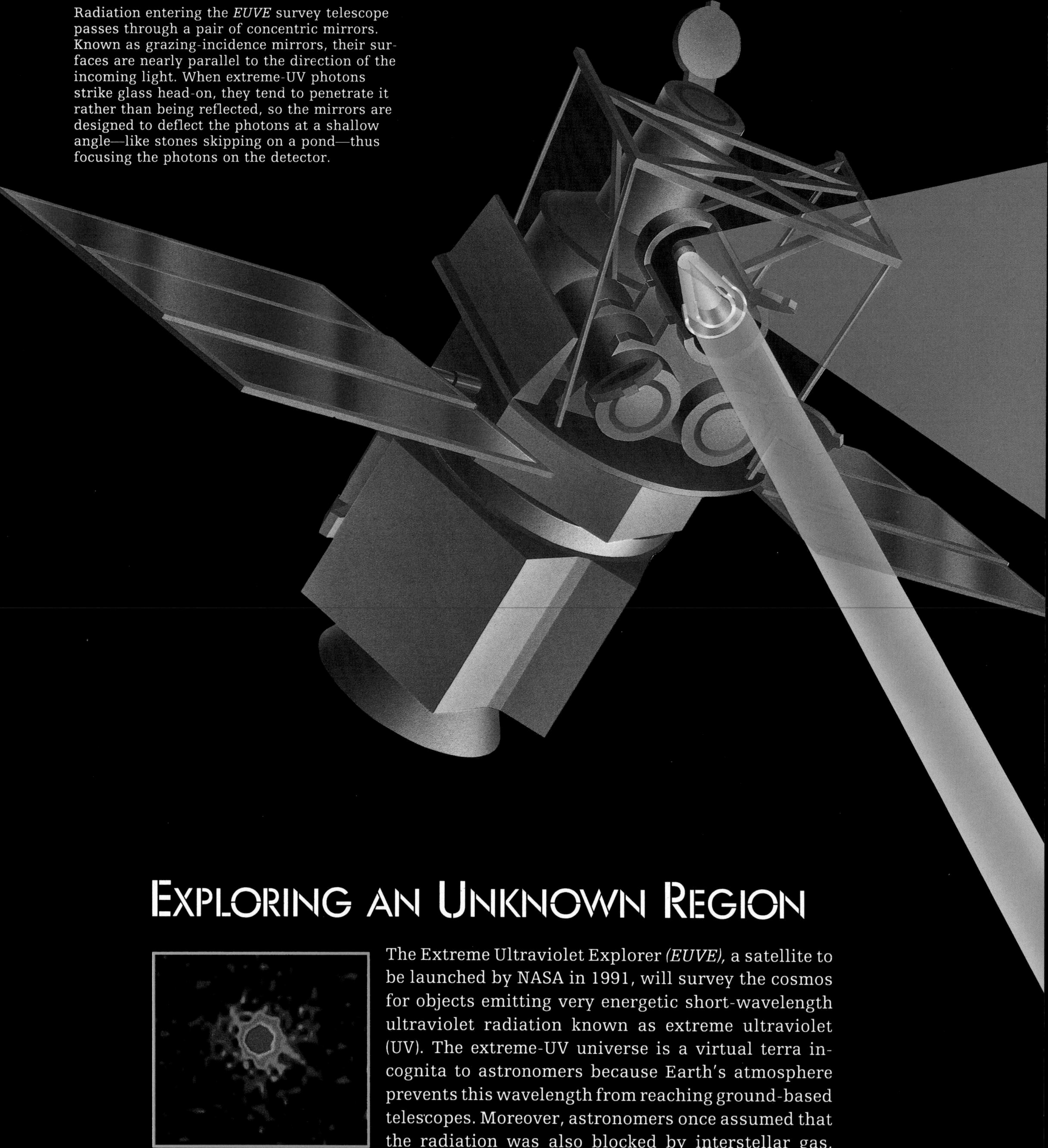

Radiation entering the *EUVE* survey telescope passes through a pair of concentric mirrors. Known as grazing-incidence mirrors, their surfaces are nearly parallel to the direction of the incoming light. When extreme-UV photons strike glass head-on, they tend to penetrate it rather than being reflected, so the mirrors are designed to deflect the photons at a shallow angle—like stones skipping on a pond—thus focusing the photons on the detector.

EXPLORING AN UNKNOWN REGION

The Extreme Ultraviolet Explorer *(EUVE)*, a satellite to be launched by NASA in 1991, will survey the cosmos for objects emitting very energetic short-wavelength ultraviolet radiation known as extreme ultraviolet (UV). The extreme-UV universe is a virtual terra incognita to astronomers because Earth's atmosphere prevents this wavelength from reaching ground-based telescopes. Moreover, astronomers once assumed that the radiation was also blocked by interstellar gas,

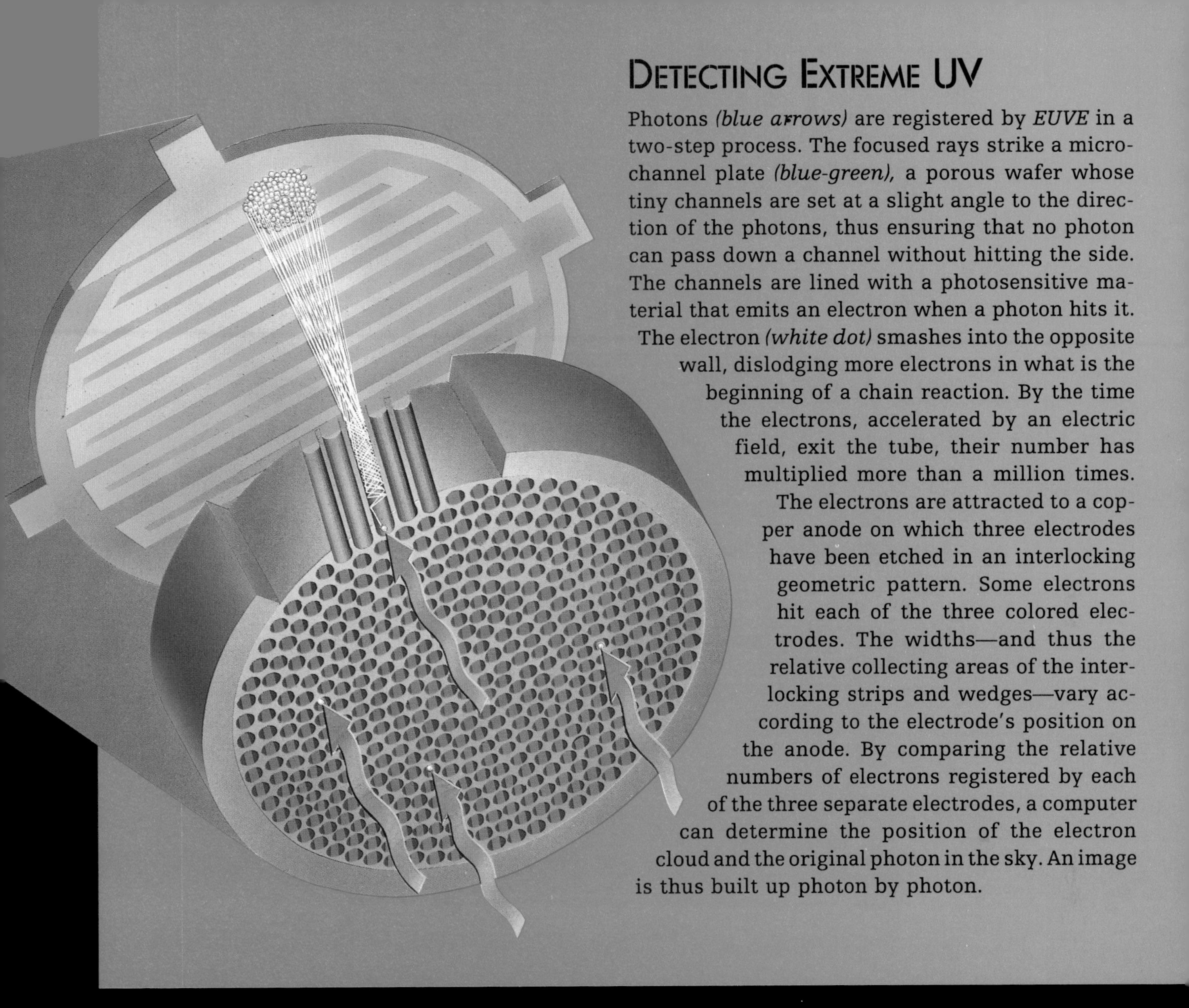

Detecting Extreme UV

Photons *(blue arrows)* are registered by *EUVE* in a two-step process. The focused rays strike a microchannel plate *(blue-green),* a porous wafer whose tiny channels are set at a slight angle to the direction of the photons, thus ensuring that no photon can pass down a channel without hitting the side. The channels are lined with a photosensitive material that emits an electron when a photon hits it. The electron *(white dot)* smashes into the opposite wall, dislodging more electrons in what is the beginning of a chain reaction. By the time the electrons, accelerated by an electric field, exit the tube, their number has multiplied more than a million times.

The electrons are attracted to a copper anode on which three electrodes have been etched in an interlocking geometric pattern. Some electrons hit each of the three colored electrodes. The widths—and thus the relative collecting areas of the interlocking strips and wedges—vary according to the electrode's position on the anode. By comparing the relative numbers of electrons registered by each of the three separate electrodes, a computer can determine the position of the electron cloud and the original photon in the sky. An image is thus built up photon by photon.

which would have made even above-the-atmosphere observing fruitless. However, that assumption was disproved in the mid-1970s, when experiments aboard the Apollo-Soyuz mission and later rocket-borne experiments detected extreme-UV radiation from stars up to 200 light-years away.

As the latest eye on the extreme-ultraviolet sky, the *EUVE* will carry a total of four instruments. Shown in detail here is a detector designed to sense the least energetic of photons in the extreme-ultraviolet range.

Because only the hottest stars are expected to emit substantial amounts of extreme-UV radiation, astronomers look forward to a bonanza of new objects from the satellite, including perhaps ten times as many white dwarf stars as are presently known. The computer-generated image at far left depicts scientists' vision of what a hot star might look like at extreme-ultraviolet wavelengths.

A Satellite for Capturing X-Rays

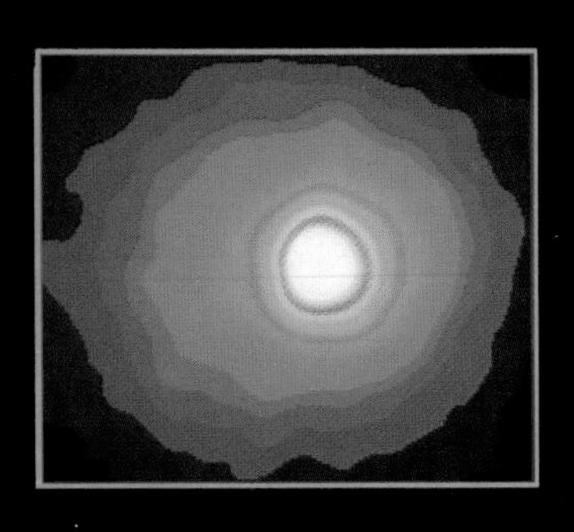

Only the most violent cosmic phenomena—such as supernova explosions or black holes—spawn x-rays, which pack a wallop 100 to 250,000 times greater than visible photons. Like extreme-UV photons, x-rays arriving at steep angles tend to penetrate mirrors rather than bouncing off. X-ray telescopes thus also use grazing-incidence mirrors but with shallower grazing angles.

The Einstein Observatory, an orbiting x-ray telescope that operated from November 1978 to April 1981, carried a variety of instruments that could be rotated into place at the focal point. One of them, shown in detail on these pages, is the imaging proportional counter, which makes pictures such as the one of the Perseus cluster at left, a vast association of galaxies some 360 million light-years away. X-rays spewing from the core of Perseus hint at explosive forces at work in that distant realm.

Astronomers are still analyzing the wealth of data returned by the Einstein satellite, a bounty that has only whetted their appetites. Planned for the 1990s is *AXAF* (Advanced X-Ray Astrophysics Facility), which will be able to detect sources 100 times fainter than those discerned by *Einstein.*

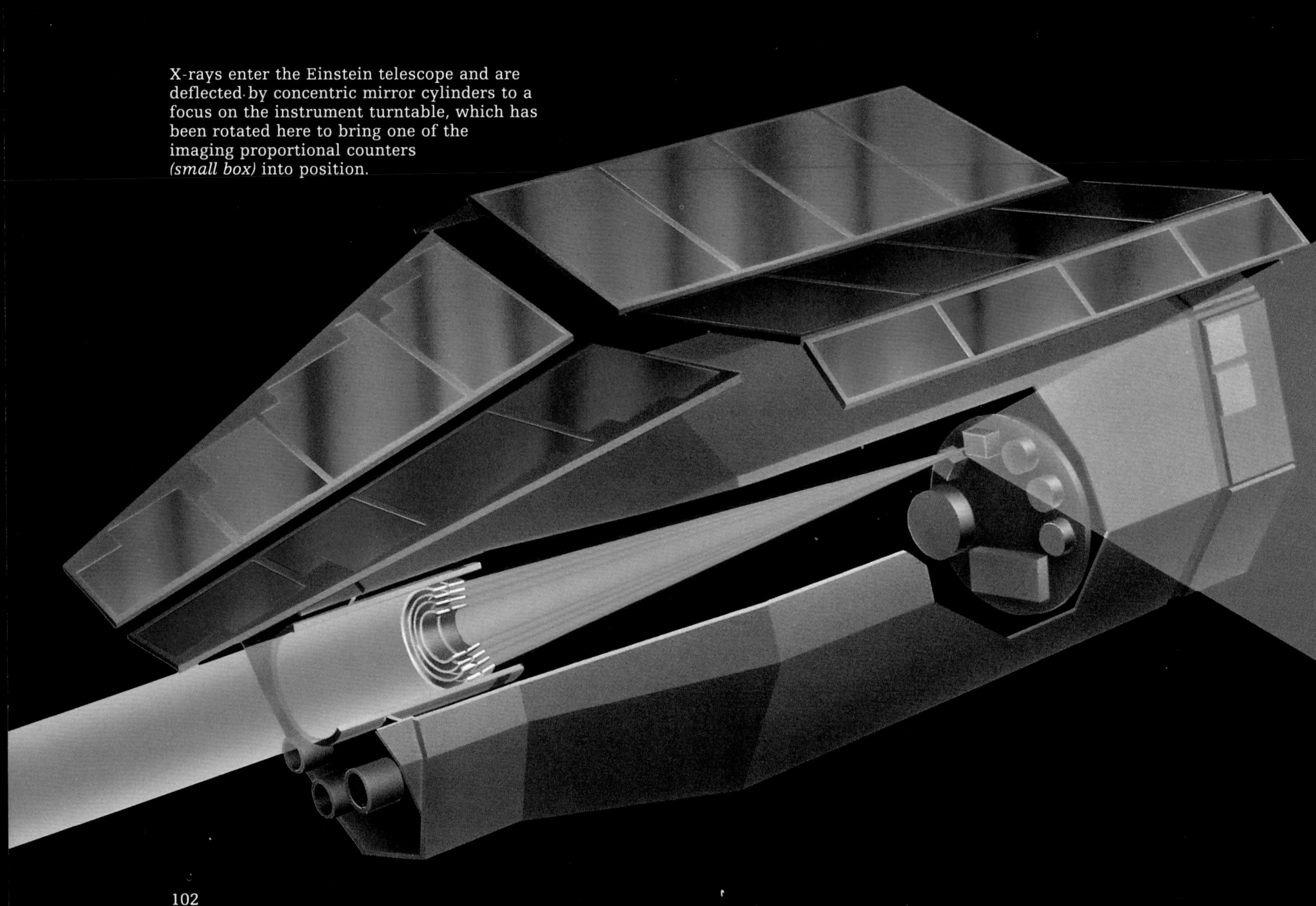

X-rays enter the Einstein telescope and are deflected by concentric mirror cylinders to a focus on the instrument turntable, which has been rotated here to bring one of the imaging proportional counters *(small box)* into position.

An X-Ray Map from Electron Tracks

The imaging proportional counter is a gas-filled chamber that is crisscrossed by three positively charged wire grids. The center anode *(yellow)* has the highest voltage and is flanked by cathodes with a lesser charge *(red).* The wires of one cathode run horizontally, and the wires of the other run vertically.

An incoming x-ray *(blue arrow)* hits an atom of the gas and knocks loose a high-speed electron *(1),* which begins knocking loose more electrons. The little cloud of electrons *(2)* is attracted toward the anode, where the cloud becomes an avalanche whose total charge is directly proportional to the energy level of the original x-ray.

When this electron avalanche occurs at the anode, it also induces *(red arrows)* an electrical signal *(3)* in each of the two cathodes. The signal moves in both directions along each wire, and its arrival at each end is timed and recorded. The difference in the arrival times of the signal at the two ends of a cathode wire indicates the place along each wire where the signal originated. Comparing these results for each cathode fixes the horizontal and vertical positions of the original x-ray. By keeping track of each photon, the IPC eventually builds up enough data to construct an x-ray image of some distant exploding galaxy or star.

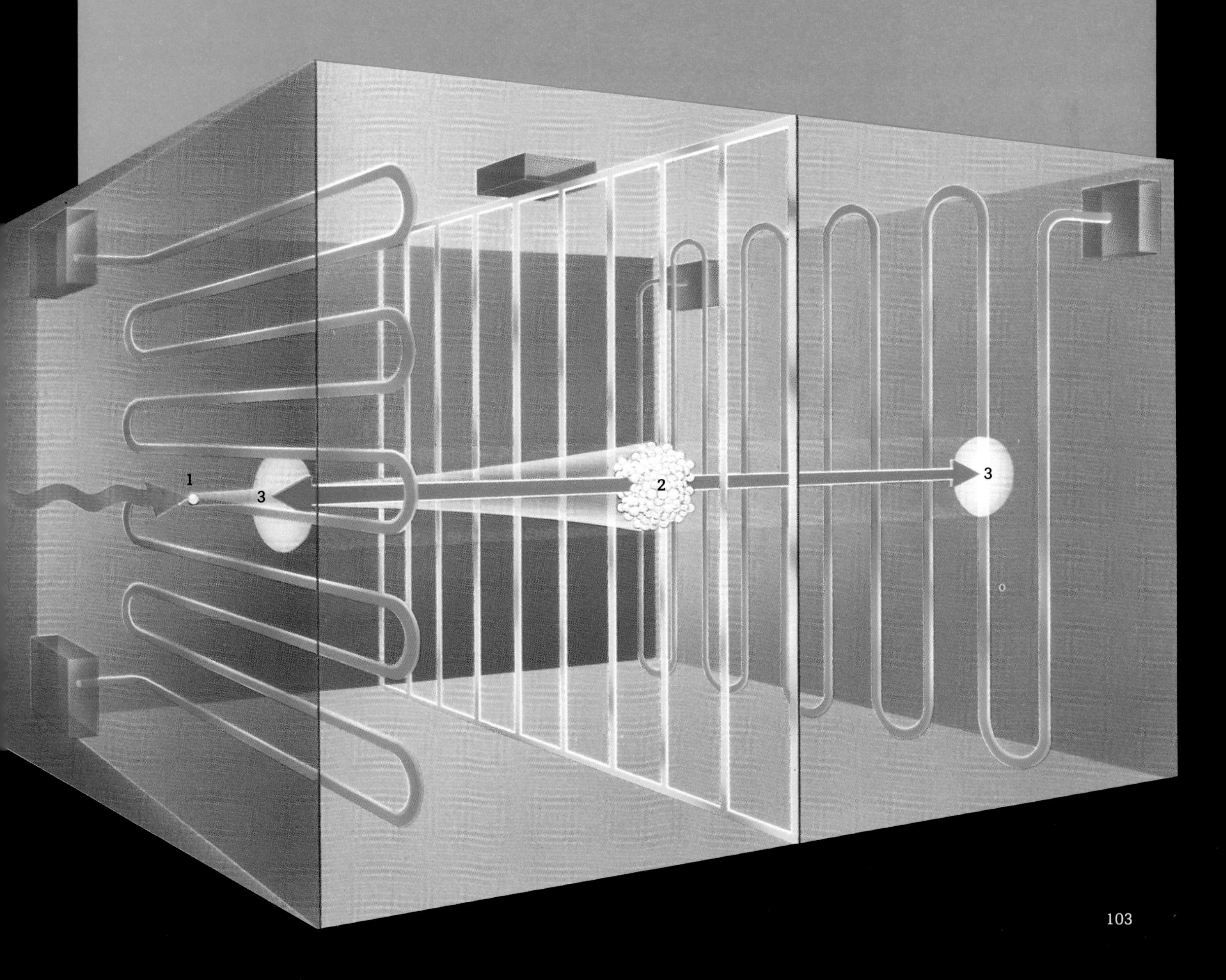

In Search of Rare Gamma Rays

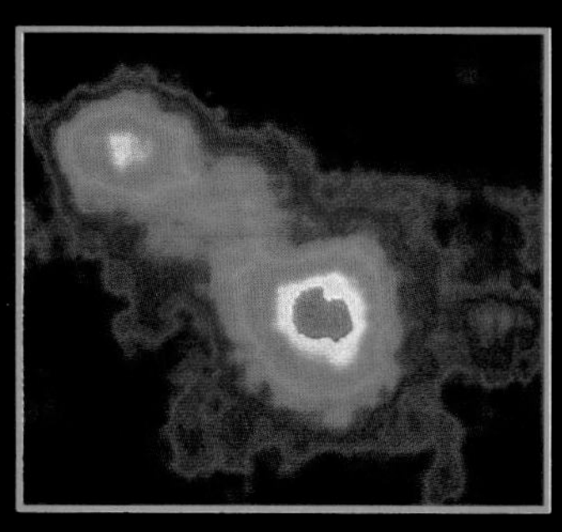

With a million or more times the energy of photons of visible light, gamma rays are by far the most energetic form of electromagnetic radiation. They are also the hardest to find. Because gamma rays are unable to get through Earth's atmosphere, they are normally observed from space, but even there they are exceedingly rare. A European satellite known as *COS-B,* for example, which produced the image of a portion of the Milky Way shown above, picked up only one or two gamma rays per hour when it was aimed at a "bright" source—and as few as seven per day when it was aimed at a faint one. (In the *COS-B* image, yellow and white denote regions of highest gamma ray intensity, blue the weakest.) In all, only about 200,000 gamma rays have ever been detected—about the number of visible-light photons received by the naked eye in one second from a star of average brightness.

Fortunately, the numbers should go up in the 1990s when NASA launches a powerful new satellite: The Gamma Ray Observatory *(GRO),* shown here, will be ten times as sensitive as *COS-B.* Among the instruments it will carry is the energetic gamma ray experiment telescope, or EGRET *(right). GRO* is unique in that its instruments will have no mirrors. Gamma rays are so energetic that they would penetrate a mirror placed at even the shallowest grazing angle. Instead EGRET takes advantage of a strange phenomenon known as pair production. According to the laws of quantum physics, a gamma ray passing close to an atomic nucleus can create a pair of particles—an electron and its antimatter partner, a positron. *GRO* astronomers use measurements of the motion and direction of the electron and positron pairs to reconstruct the energy and direction of each gamma ray, thereby building up a picture of a gamma ray source.

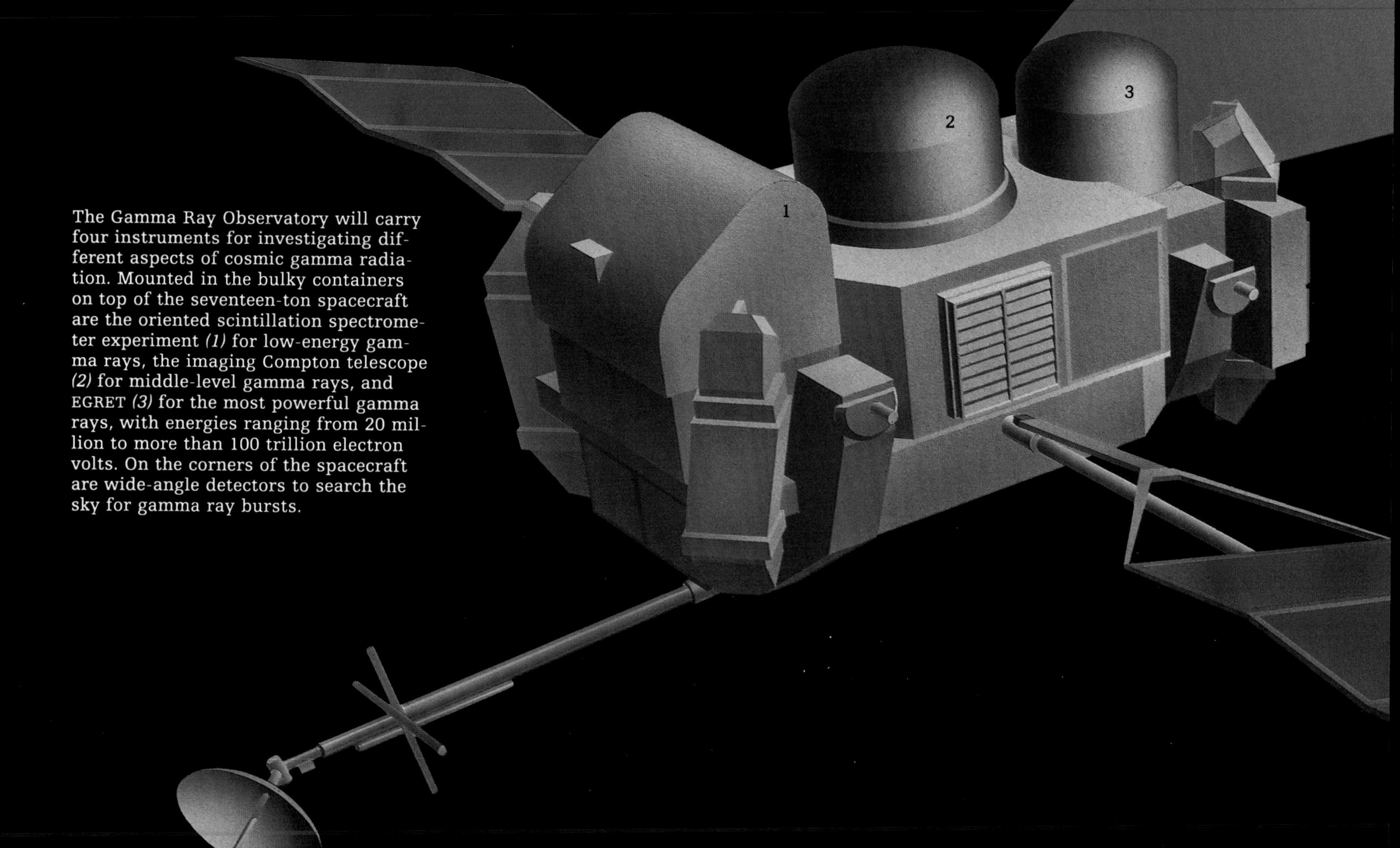

The Gamma Ray Observatory will carry four instruments for investigating different aspects of cosmic gamma radiation. Mounted in the bulky containers on top of the seventeen-ton spacecraft are the oriented scintillation spectrometer experiment *(1)* for low-energy gamma rays, the imaging Compton telescope *(2)* for middle-level gamma rays, and EGRET *(3)* for the most powerful gamma rays, with energies ranging from 20 million to more than 100 trillion electron volts. On the corners of the spacecraft are wide-angle detectors to search the sky for gamma ray bursts.

A Sparkling Trail

EGRET employs the technology of high-energy particle physics to take photographs of gamma ray sources. When a gamma ray *(blue arrow)* enters the gas-filled EGRET chamber, it plows through twenty-eight narrowly spaced spark chambers made of tungsten plates *(1)*, transforming itself into an electron-positron pair traveling at nearly the speed of light. The two particles shoot down through a scintillator *(2)*, which emits a flash of light when particles traverse it. The particles continue through a series of widely spaced spark chambers *(3)* and then pass through another scintillator *(4)*. The scintillators are connected to a special circuit that determines—in less than a millionth of a second—that a particle has passed. The circuit causes a high voltage to be applied across two sets of wires in each spark chamber, causing sparks to fly along the ionized track left in the gas by the particles. The particles' path reveals the direction of the incoming gamma ray. Finally, the particles hit a layer of sodium iodide crystal, triggering an avalanche of electrons and a burst of light *(5)*, measured by a bank of photomultipliers *(6)*. The amount of light tells scientists the energy of the electron-positron pair, from which they can calculate the energy of the original gamma ray.

3/APPROXIMATIONS OF TIME,

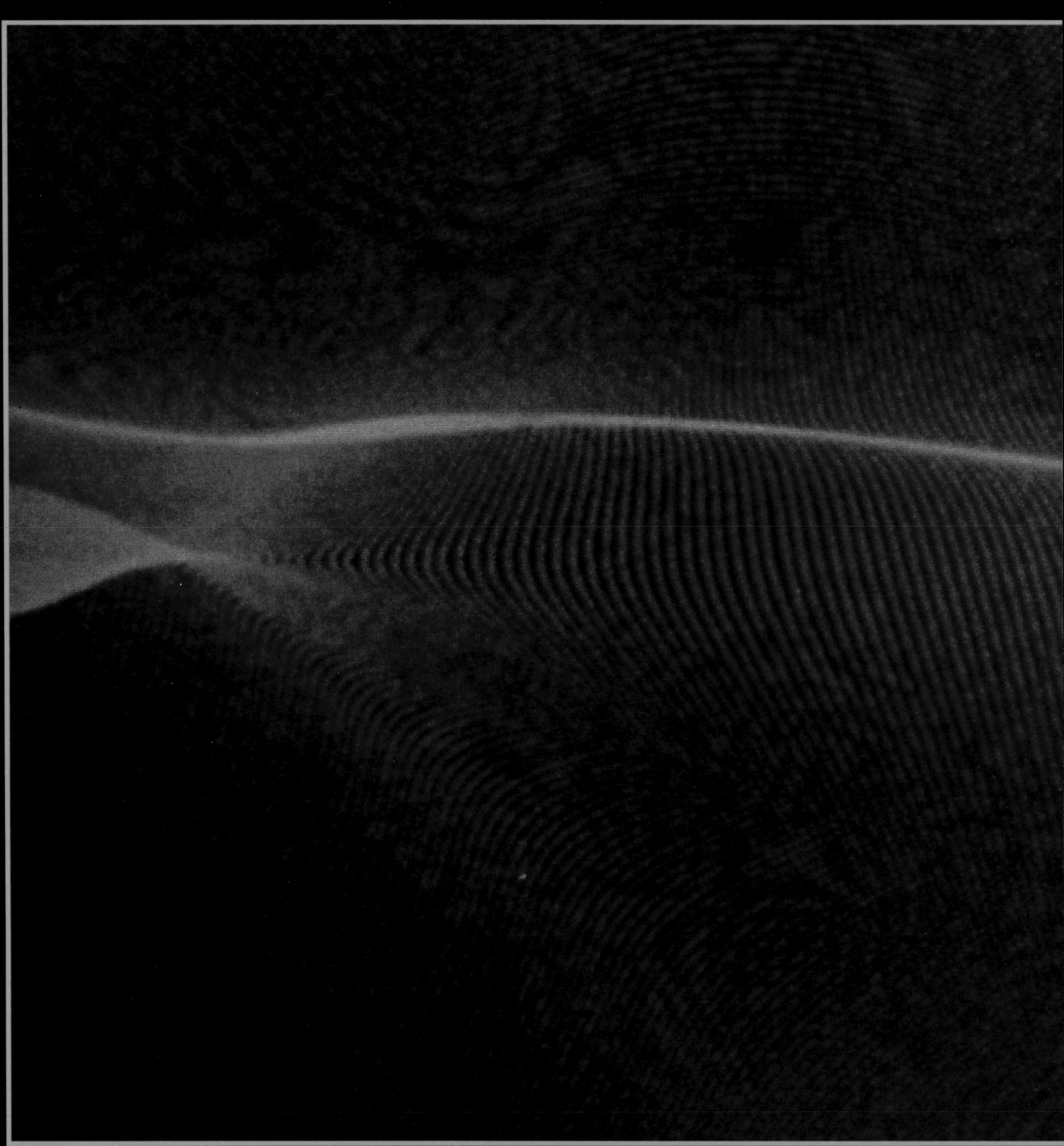

SPACE, AND MATTER

The cosmos evolves in a computer, with imaginary galaxies gathering in clumps and bridges. The structure displayed in this model by Adrian L. Melott of the University of Kansas is remarkably similar to that observed in the real universe.

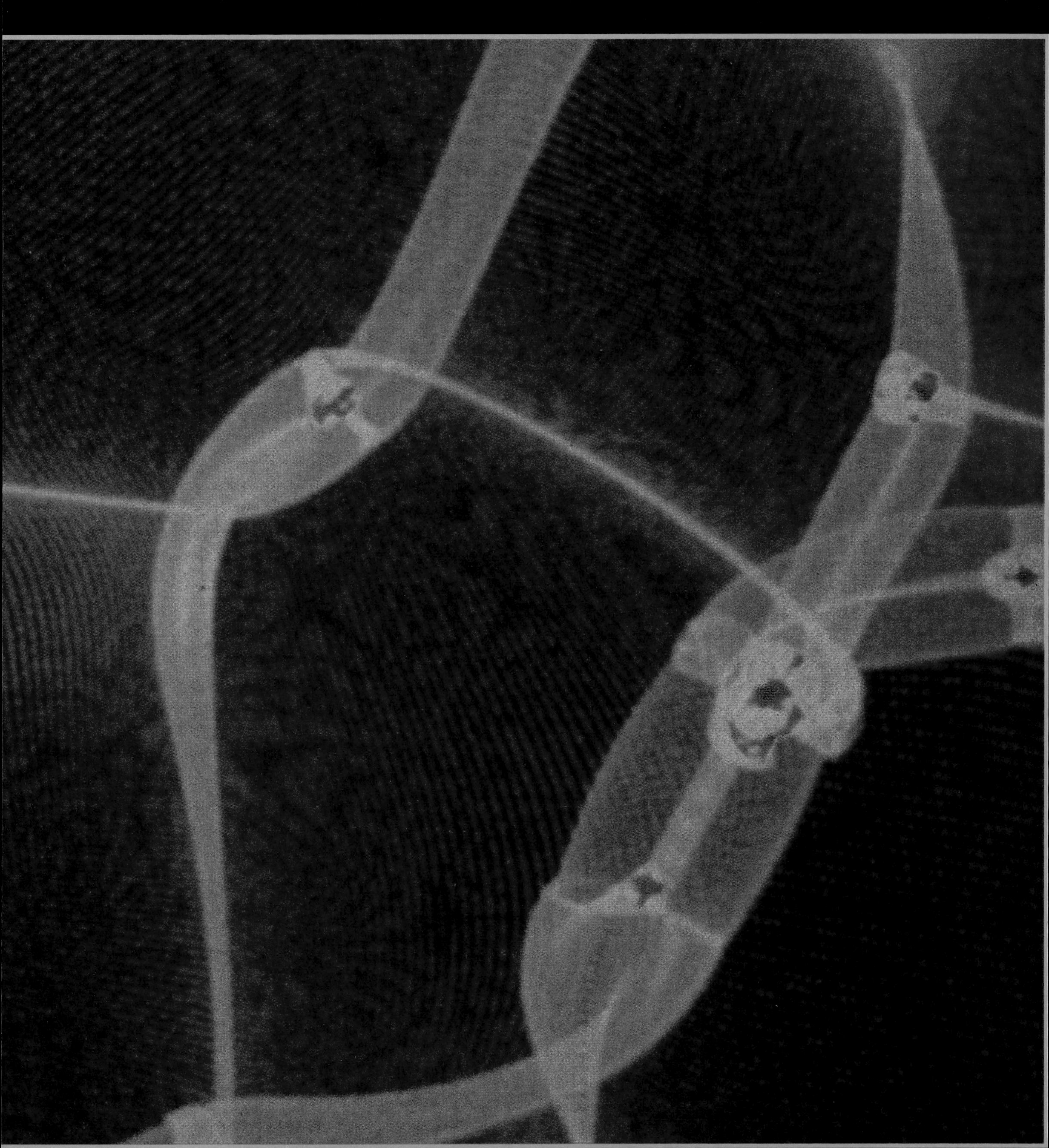

wedish astronomer Erik Holmberg must have felt a surge of pride as he watched two swirling galaxies sideswipe each other in his lab at the Lund Observatory in 1941. Scarcely a year earlier, he had conjectured that a pair of galaxies could deform one another as they passed in space, and now—in the complex arrangement of movable lights before him—the thirty-three-year-old scientist possessed a model that brought his idea to life.

Flouting the astronomical wisdom of the day, which held that galaxies form in splendid isolation, Holmberg suggested that a galaxy can take shape in response to the fierce gravitational forces of other galaxies passing nearby. No galaxy is an "island universe," he insisted; instead, each is apt to collide eventually with its neighbors, inflicting cosmic scars that linger briefly in the form of spiral arms.

To support his view, Holmberg crafted a laboratory model capable of reenacting the stages of these extraordinary pas de deux. He knew it was possible to calculate the gravitational effects (called tides for their tendency to ebb and flow) that arise when one galaxy crosses the path of another, but in the precomputer days of the early 1940s, the volume of pencil-and-paper mathematics required to produce an answer made the exercise impractical. However, as he reported in the November 1941 issue of the *Astrophysical Journal,* "a solution has been found by replacing gravitation by light."

The validity of such a substitution—indeed, the success of the entire simulacrum—hinged on the fact that gravity and light diminish to the same degree over a given distance, a property described by the inverse-square law of physics. Holmberg therefore stood on firm scientific ground as he represented the myriad stars of a hypothetical galaxy in the simplified form of thirty-seven tiny light bulbs arrayed in four concentric rings. The brightness of the bulbs, which had been specially designed and manufactured so that their intensity would hardly vary, stood for the mass—and therefore the gravitational pull—of the stars in the system.

To achieve an accurate distribution of mass throughout his model, Holmberg extrapolated from stellar observations showing that the surface brightness of a galaxy rises rapidly toward its center. He attained a similar effect by graduating the voltage supplied to each ring of lights: The inner two circles burned with a candlepower of 1, while the third ring burned at .7 and the outer ring at .3. He then fashioned a second mock galaxy the same way.

Mounted on a table covered by a grid, the two sets of custom-made bulbs could be inched along in step-by-step mimicry of what Holmberg imagined must occur when a pair of galaxies pass in space. As those movements were made, an array of photocells measured the minute fluctuations in the amount of light reaching the boundary of the grid; the photocells translated these optical signals into electrical signals and forwarded them to a second bank of instruments known as galvanometers, which charted the rising and falling current levels with precision. In theory, the variations in these readings would mirror the gravitational tides that course through galaxies in collision.

Holmberg's experiments led him to conclude that, given the proper conditions, one star system could capture another during their intergalactic dance. The gravitational tides created by the mutual tugging of the two galaxies would bleed energy from their orbital motions and force them into ever-smaller, ever-closer orbits.

THE VISIONARY SPURNED

Although Holmberg's basic premise has since been accepted and elaborated by astronomers, his methods won few accolades at the time. Astronomy reigned as the purest of the observational sciences in the 1940s, and a number of its practitioners resisted the notion that large-scale celestial phenomena could be studied by proxy on Earth. Biologists and chemists might cook up approximations of nature in their petri dishes and beakers, ran this argument, but astronomers could only wait patiently for the universe to divulge news of events that had occurred millions of years ago and light-years away.

More than two decades would pass before astronomers recognized Holmberg's creation as the forerunner of modern-day astronomical modeling. Today, powerful computers programmed with the rules governing matter and energy stage mock versions of stellar, galactic, and cosmic happenings. Astronomers prize such simulations because they yield insights about the workings of the universe that cannot be obtained by pure observation alone. Computer models thus sweep aside the obstacles of both space and time: Astronomers can not only re-create unimaginably distant events in laboratories close at hand but also toy with the history of the heavens, watching how the most minute alterations in a system's makeup produce repercussions that cascade through time. For these reasons, computer modeling has become a prime investigative tool of theoretical astrophysicists and a mainstay of the new astronomy.

As Holmberg did in 1941, many modelers today conceive of astronomical bodies as collections of points. The points can represent any type of variable, from mass (signified by Holmberg as unwavering sources of light) to temperature to elemental composition. Despite these similarities in subject and approach, the tabletop model and its computerized descendants differ radically in complexity and abstraction. Where Holmberg's version consisted of a mere seventy-four hand-movable points, for example, astronomical models of the 1980s were run on supercomputers and could track the velocities and positions of as many as a million imaginary particles. The latest models incorporated as many as 10 million equations, all intertwined to reveal how each variable would change from its given starting condition.

Solving such models is an iterative process, and the model builder is free to dictate how much cosmological time each iteration shall represent. If a single run—that is, a single solution of all the model's constituent equations—covers the passage of 1,000 years, then several thousand iterations may suffice to re-create, stepwise fashion, the evolution of a heavenly body or process over thousands, millions, or billions of years. During the 1960s, when hardware designers first produced machines equal to the task, astronomers began to relegate the drudgery of number crunching to digital computers. Since then, computer models have been used to examine or explain all sorts of elusive phenomena, from binary stars, neutron stars, and flaring starspots to quasars, galactic jets, and the life cycles of galaxies.

STELLAR EQUATIONS

The appeal of computer models to astronomers today can be gauged by examining how their counterparts fared without them in the past. Among the first sky watchers to attempt reducing the process of stellar evolution to a set

A Model Universe

The advent of supercomputers gave astronomers a way to test their ideas about the universe by simulating phenomena on a vastly reduced time scale. With computer models, scientists can create snapshots of cosmic events at intervals (called time steps) of several million years. In "particle models," shown here and on pages 112-113, each star or galaxy is treated as a particle of matter concentrated at a single point, with a given mass, density, and velocity. The computer calculates the gravitational influence of the particles on one another for one time step, then moves each particle accordingly. After recording the new positions, the computer repeats the process for another time step. Since more particles require more computing, astronomers use statistical techniques to reduce their number, making up for slight losses in accuracy with substantially reduced calculating time.

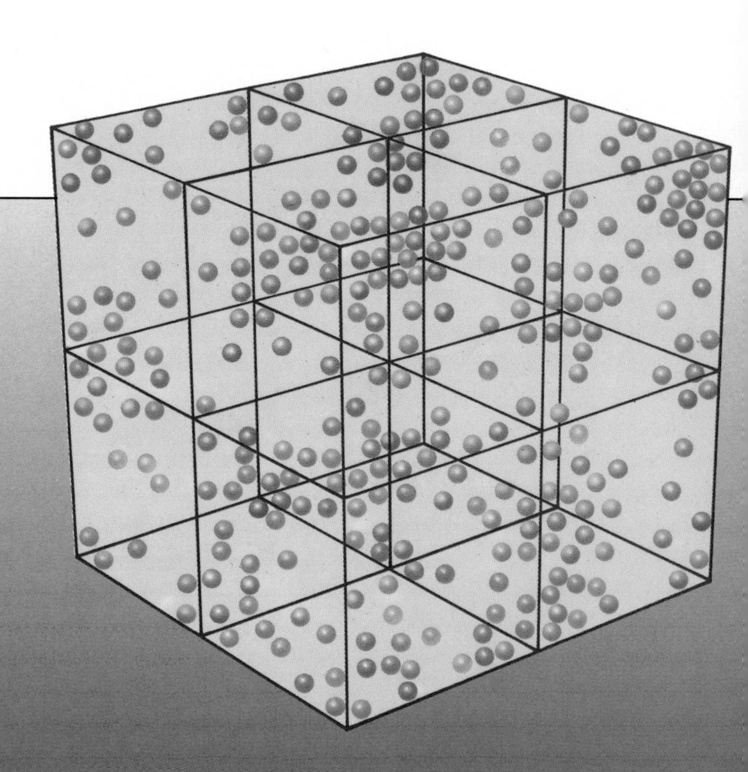

In the imaginary space of a particle-mesh model, a three-dimensional grid is superimposed on the particles to be studied (here bodies of equal mass, randomly distributed). The grid for a typical model may contain 250,000 cubes, or cells; for clarity, only eight are illustrated here.

of mathematical equations was Arthur Stanley Eddington, the dean of British astronomy and physics during the 1920s and 1930s. In 1924, Eddington produced a model showing that as the internal temperature of certain stars reaches many tens of millions of degrees Kelvin, the resulting radiation pressure can rip the stars to gaseous shreds. Yet even that effort required a prodigious number of calculations, and further progress with the model came only in fits and starts.

The appearance of computers on the scientific scene after World War II therefore tantalized astronomers no end. These early vacuum-tube machines were prone to breakdown, but they offered welcome relief from the tedium of constructing models by hand. The relief was not immediately available to everyone, however; the first computers resided in a few select labs worldwide, where their use was strictly controlled.

Perhaps the first astrophysicist to put the mathematical marvels to extensive use was Princeton's Martin Schwarzschild. In 1948, Schwarzschild, who had fled the Nazi regime in his native Germany in 1936, began developing four interlocking equations that he believed encapsulated the process of stellar evolution. He programmed them into a prototype computer built by mathematician (and fellow Princetonian) John von Neumann. As the quartet of equations ran on von Neumann's machine, they computed the brightness,

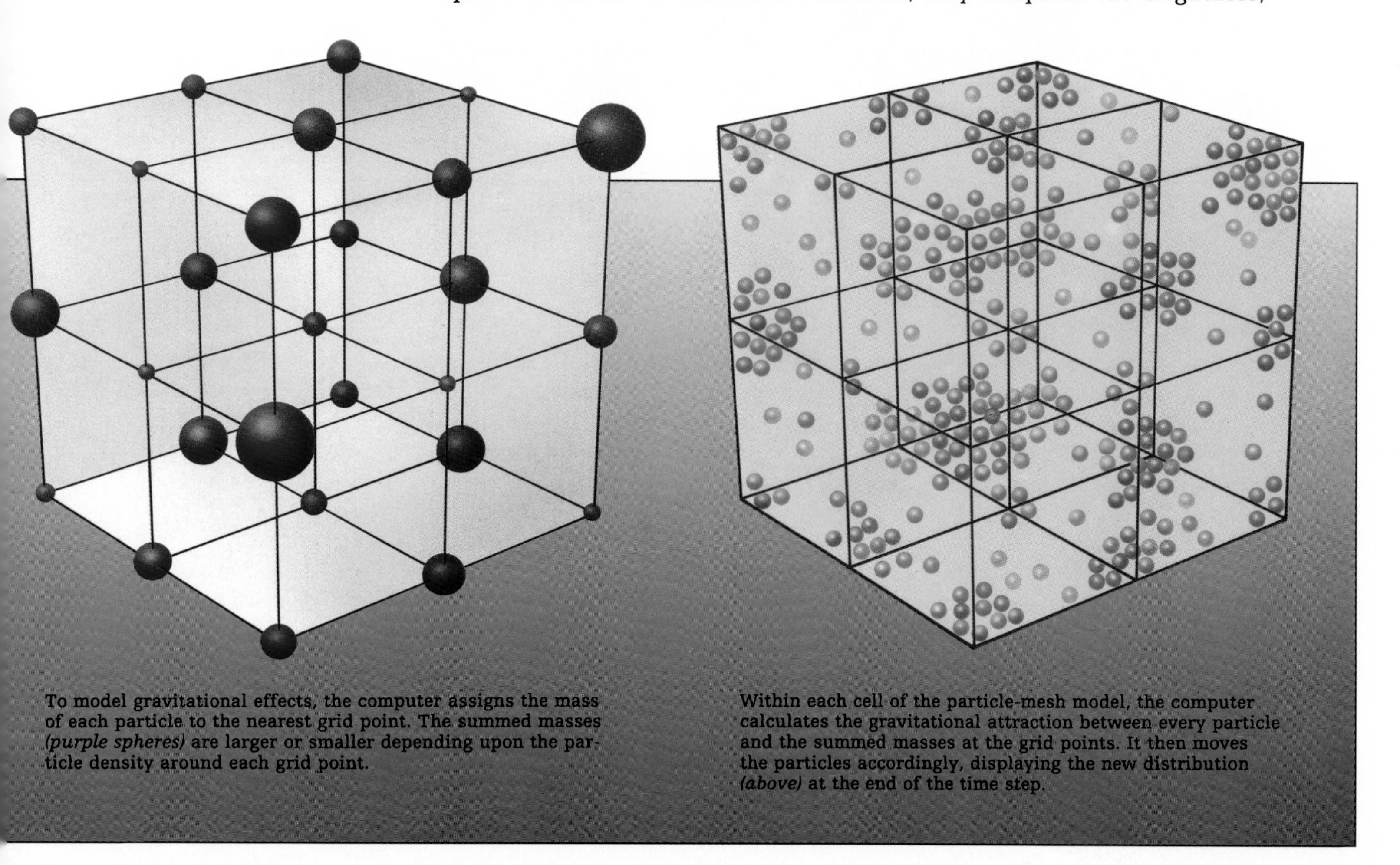

To model gravitational effects, the computer assigns the mass of each particle to the nearest grid point. The summed masses *(purple spheres)* are larger or smaller depending upon the particle density around each grid point.

Within each cell of the particle-mesh model, the computer calculates the gravitational attraction between every particle and the summed masses at the grid points. It then moves the particles accordingly, displaying the new distribution *(above)* at the end of the time step.

temperature, mass, and chemical composition of a hypothetical star over the course of millions of years. The machine's malfunctioning vacuum tubes made the process a frustrating one, yet Schwarzschild's model yielded tangible results: It revealed that stars transform in surprising ways as they age, switching from hydrogen to heavier nuclear fuels and ultimately ballooning into gigantic, reddened versions of their former selves.

One of Schwarzschild's compatriots, the German astrophysicist Rudolf Kippenhahn, also pioneered in harnessing the might of the new machine. At the Max Planck Institute for Physics in Göttingen in 1957, Kippenhahn and a team of fellow researchers used a sprawling, room-size computer called the G-2 to simulate stellar aging. Sweltering in the heat thrown off by the machine's vacuum tubes, the researchers tensely monitored the running of their programs, Kippenhahn later recalled, praying "that the computer would continue to work for five hours without breaking down, else we would have to begin all over again."

As cooler and more reliable transistors replaced vacuum tubes in the early 1960s, perspiration and supplication receded from the computing arena. The machines' high cost, however, remained a barrier to widespread use. Because the U.S. military was among the handful of institutions that could afford them, transistor-based mainframes, as large computers are known, soon found a niche in Defense Department research labs around the country.

At one of these national labs, located in Los Alamos, New Mexico, IBM's new 7030 mainframe—dubbed Stretch because its reach so far exceeded that of earlier machines—was put to work in April 1961 on the numerical analysis

The Accuracy of Particle Trees

A recently developed technique called particle-tree modeling, shown in simplified form at right, often produces more accurate results than particle-mesh simulations, with about the same computing effort. Instead of overlaying its imaginary space with a uniform grid, a particle-tree model cuts the space into several large cubical cells. (Cubes are represented here as squares for clarity.) A cell containing more than one particle is further divided into subcells, each half as long, high, and wide. Any subcell with more than one particle is subdivided in the same way, until the entire space is broken into cells and subcells of many sizes, each either empty or containing one particle.

In the example shown here, particles of equal mass are randomly distributed on the bottom layer, which has been divided into cells sized so that none holds more than one particle. So-called pseudoparticles in the cells of the upper layers summarize the characteristics of the particles in the squares immediately below them. In this case the distribution of mass in the bottom layer is reflected in the varying sizes of the pseudoparticles. A single pseudoparticle at the top sums up all the masses below it in the hierarchy. An actual model may have as many as 100,000 particles and require up to a dozen levels.

A two-part calculation finds the gravitational force on each particle. First the computer works out the interactions with neighboring particles, one at a time. Then it adds in the effect of more distant particles, considering them as groups, represented by the appropriate pseudoparticles from the next level of the tree. As the distance increases, the computer calculates interactions with pseudoparticles from higher levels, which stand for larger groups.

of thermonuclear explosions. Computer models of such detonations have fostered the hand-in-hand growth of supercomputers and nuclear-weapons technology since the mid-1960s because the models are critical to the design of ever-deadlier hydrogen bombs. What drew astronomers to the simulations was their re-creation of fusion, a momentary destructive force in nuclear blasts but a lifelong creator of energy in stars. Nuclear physicists, intrigued by this larger role of fusion, began writing programs of their own to reconstruct the cataclysmic events surrounding the most powerful explosion of all—that of a star into a supernova.

SIMULATIONS COME OF AGE

One government researcher much inspired by these investigations was Jim Wilson. In 1968, Wilson set out to explore the mechanics of supernovae and to examine the recently discovered rotating magnetic stars known as pulsars. One of Wilson's innovations, a sophisticated scheme for displaying the data delivered by his models, presaged the sort of realistic effects that computer graphics would make available to modelers in the next two decades.

But it was Wilson's simulation of a star's collapse in 1970 that helped computer models come of age in astronomy. To speed the running of his model, Wilson treated the star as a fluid in motion. In this approach to stellar studies, called hydrodynamics, the gas that makes up a star is assumed to obey the same behavioral laws as a liquid. The number-crunching aid of a computer is almost essential to the method. Using a supercomputer at the Lawrence Radiation Laboratory (renamed the Lawrence Livermore Laboratory a year later), Wilson calculated how the rotation and magnetic fields of a dying star might trigger key changes in the temperature, density, and pressure of its constituent gases.

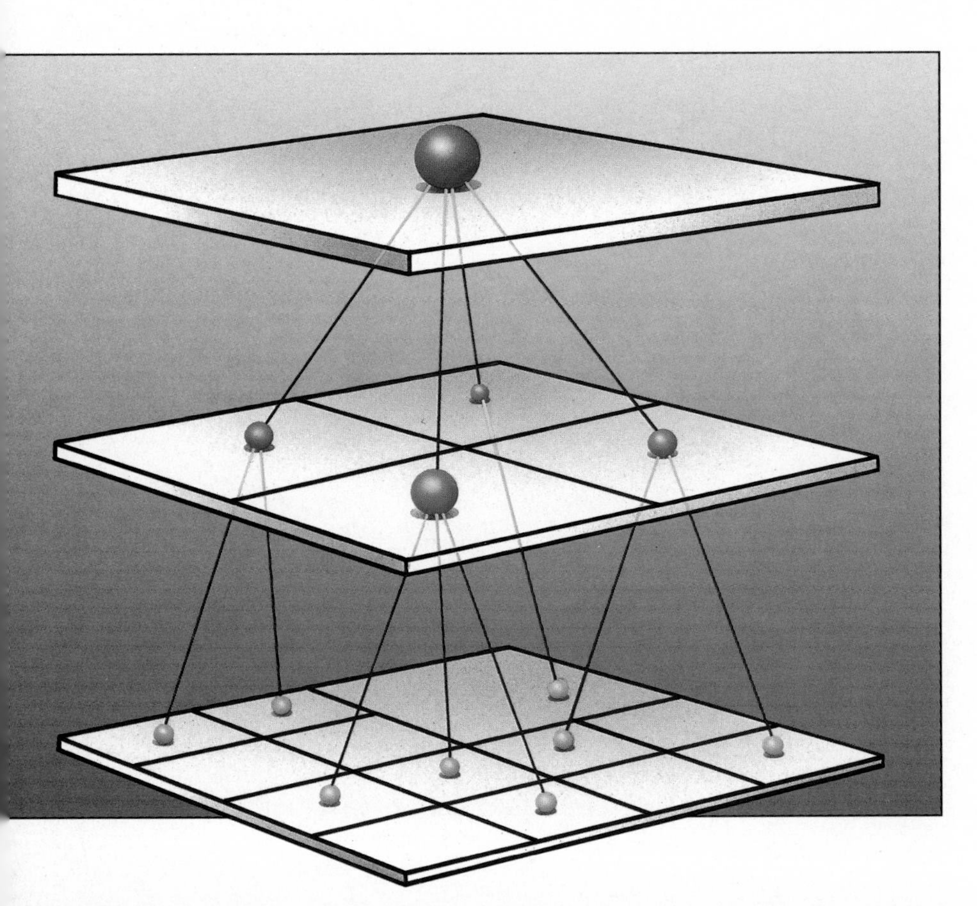

The stellar aging process was so complex that Wilson had to limit the number of calculations in his model, a handicap that yielded less-than-ideal results. Crude as they were, however, such results were unquestionably superior to those produced by slide rules and logarithm tables. Astrophysicist Icko Iben, Jr., of the University of Illinois, a pioneer of fluid modeling in the 1960s, once estimated that the calculations necessary to model the life cycle of a star would take an astronomer 3,000 years to complete by hand, whereas the same process could be simulated on a computer in a matter of eight hours.

Access to computers remained the central difficulty for a number of years. Throughout the 1970s, Iben made jocular but genuine pleas in the scientific media on behalf of astronomers, claiming that their "mental, emotional, and physical health" suffered from the constant hardship of cadging computer time from engineers, computer scientists, and others who had easier access to the machines. By the end of the decade, Iben's appeals had swayed a few government and industry laboratories. As astrophysicists gained more time on the machines, computer modeling of the heavens made some great leaps forward.

TRAFFIC ACCIDENTS IN THE UNIVERSE

In the forefront of the computer investigations were two Estonian refugee brothers, Alar and Juri Toomre, who had arrived in the United States in 1949 when Alar was thirteen and Juri was ten. "We were never particularly close as kids," the elder Toomre recalled forty years later, "but both of us drifted into similar scientific things." Both Toomres majored in aeronautical engineering at the Massachusetts Institute of Technology because, as Alar put it, "physics is wider and more glamorous, but most immigrant kids want job security. Engineering is secure." In the end, however, glamour would win out over security.

After receiving their doctoral degrees from British universities, the Toomres returned to the northeastern United States to teach, Alar in Cambridge at MIT and Juri at New York University. In addition to his classroom

THE FLUID COSMOS

Many celestial objects are turbulent masses of gas or plasma rather than discrete particles. Scientists who simulate such phenomena as galactic jets, the solar wind, and Earth's own magnetotail *(pages 118-119)* usually employ a technique known as fluid modeling. This method relies on the principles of fluid dynamics—the same principles long used to describe terrestrial processes such as currents and eddies in rivers.

In a fluid model, the computer divides a space into cells, then analyzes the matter within each cell in terms of such variables as pressure, temperature, and flow velocity. Cell-by-cell changes in each variable are plotted as changes in the amplitude of a complex wave *(near right, top)*. Using a mathematical tool called Fourier analysis, the computer reduces the complex wave to its simplest components, shown combined and separately *(near right)*. Each of these so-called sinusoidal waves has a characteristic wavelength and amplitude.

The computer then applies the laws of fluid dynamics to calculate the interactions among these simple waves and the effects of other variables on them, working out their evolution for a single time step. The result is a new set of sinusoidal waves, all with their original wavelengths but some with new amplitudes *(far right)*. These waves are mathematically reassembled to create a new complex wave *(far right, top)* that describes the rise and fall of the variable across the entire space at the end of the time step.

Even a relatively simple fluid model may contain three or more variables and a grid of 10,000 cells. A supercomputer can run models with as many as a million cells but requires many hours of intricate calculating to step through the imaginary eons.

duties, the younger Toomre plunged into astrophysical research at a local NASA facility, the Goddard Institute for Space Studies. The two brothers then began a scientific collaboration in which they subjected each other's results to unusually intense scrutiny. "Brothers are almost like a husband-and-wife team," observed Alar in 1989, "excessively frank with each other. You know, it's 'What! You dummy, you wrote that?' "

The Toomres became captivated by a set of galactic deformations that would have made modeling pioneer Erik Holmberg feel right at home. Back in 1956, Fritz Zwicky of the California Institute of Technology had alerted astronomers to their existence; he named the misshapen objects "peculiar galaxies" for their failure to conform to the symmetry and regularity of 98 percent of the known star systems.

Although the cosmic curiosities went largely unstudied for the next decade and a half, they were not totally ignored. During the late 1950s and early 1960s, two astronomers, the Russian Boris Vorontsov-Velyaminov and the American Halton Arp, kept instruments focused on the strange bodies. Between them they managed to catalog more than 300 systems

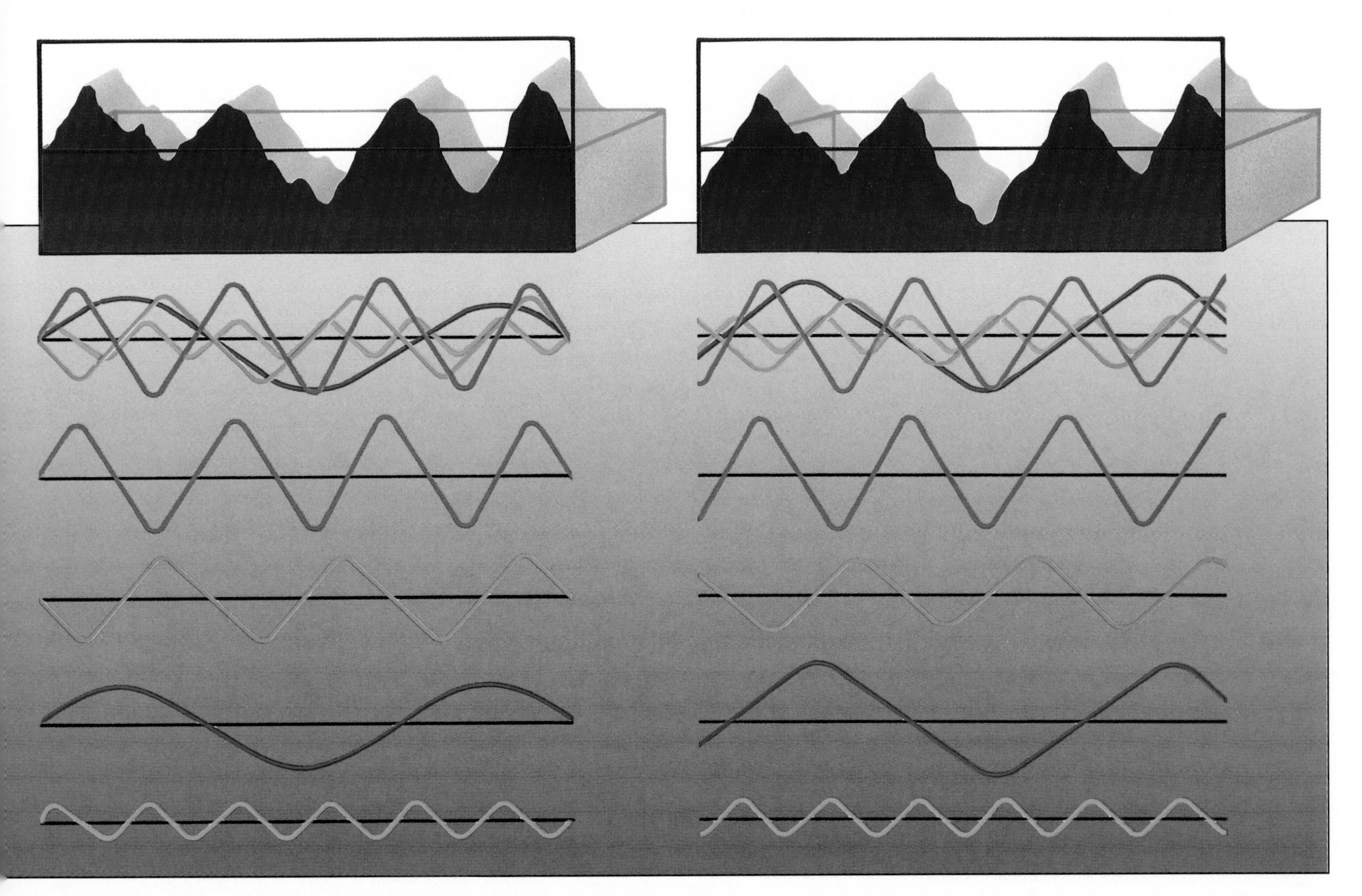

that displayed what Zwicky had identified as the aftereffects of a galactic collision or near miss: long, tail-like appendages or interposing stellar bridges. Arp was so taken with what he described as the "twisted, distorted shapes and curious linkages" that he released his findings as the *Atlas of Peculiar Galaxies* in 1966.

Having attracted the attention of Zwicky, Vorontsov-Velyaminov, and Arp, peculiar galaxies now began to exert their pull on Alar and Juri Toomre. "Almost every crowd includes a few charming eccentrics or confounded exceptions," the brothers wrote of their work. "This is true of the 'crowd' of galaxies." Because Juri enjoyed access to Goddard's IBM mainframe through his astronomical acquaintances, he and Alar chose that computer to model the gravitational interactions of pairs of passing galaxies. Juri would remain at NYU to supervise the computer's running of the model, while Alar made periodic visits from MIT to fine-tune the simulation.

The brothers simplified the galaxies in their model by representing each one as a disk whose mass was concentrated in a central sphere; each galaxy's tens of billions of stars were in turn represented by 350 massless points clustered around this high-mass core. The decision to localize each galaxy's mass at a single point in its center was motivated by some high physics and some lowly computer hardware.

First the physics: The brothers were well aware that in a real galaxy, mass is dispersed over the disk. However, drawing on the work of German astronomer Jörg Pfleiderer, the Toomres had concluded that the outer parts of a galactic disk were relatively low in mass and therefore weak in self-gravity, or the amount of gravity they exerted on one another. Their model of intergalactic traffic accidents, they reasoned, would thus "remain basically valid even if the peripheral mass is neglected entirely."

The second incentive for simplification was the limited processing power of the day. A completely realistic model would have required the Goddard computer to calculate the gravitational pull that every particle in two interacting galaxies—a total of 700 particles—exercised on the motion of every other particle for each time step in the simulation, a dictate that would have demanded a hundredfold increase in processing time. Instead, the Toomres arranged for the motion of each particle to be influenced by just two forces, represented by the central masses of the two model galaxies. "All the other particles were just going along for the ride," explained Alar Toomre in 1989, "and that made the computation vastly easier."

By changing the model's parameters, or conditions, before each run, the brothers managed to simulate all manner of crash scenes: The two colliding galaxies could be made the same size, or one of the galaxies could dwarf its partner. The systems could sail toward each other with the planes of their orbits at different angles, or they could cross paths while rotating in opposite directions. For every set of parameters, a device known as a Stromberg-Carlson plotter turned out diagrams with alphanumeric labels signifying each particle's distance from its original galactic center. When the diagrams emerged, Juri transmitted them to Alar in Cambridge using the early 1970s forerunner of the facsimile machine, the Greyhound bus.

As the brothers gained sufficient confidence in their model to publicize its output, Alar meticulously retraced dozens of diagrams by hand, replacing the plotter's typographic soup of particles with neat circles. The Toomres' paper, published in 1972 as "Galactic Bridges and Tails," showed that when spiral galaxies rotating in the same direction approach each other in slow orbits, gravity can smear the stars in each galaxy across the canvas of deep space in flowing strokes. As a small galaxy grazes a larger one, the Toomres demonstrated, a bridge of stars often forms, and when two galaxies of equal size meet, each is likely to sprout a stellar tail. This second finding apparently explained the formation of two of the most intriguing galactic pairs in Arp's atlas, those nicknamed the Mice and the Antennae. The Mice—officially designated NGC 4676A and NGC 4676B—were so titled for the long tails that stretch behind them into space; the Antennae display a similar structure, with opposing tails that arch like the tentacles of an insect.

Most revealing of all, the Toomres' computer model demonstrated that many perceived features of interacting galaxies are in fact optical illusions. Earthbound observers are confined to a single vantage point on the cosmos, but advanced computer graphics allow modelers to adjust their angle of view. These new perspectives have exposed the true shapes of a number of galactic features that have puzzled astronomers over the years. The Toomres' model showed, for example, that the thin filament seen linking M51, also known as the Whirlpool galaxy, to its neighbor NGC 5195 is not a galactic bridge at all but a so-called projection effect caused by the two galaxies' relative positions along the line of sight from Earth. As Alar Toomre noted after the discovery, with computers "you can reach things you can't possibly touch by hand."

BRAVE NEW WHORLS

The growth of computer speed and power throughout the 1970s and 1980s made possible ever-more-ambitious simulations. Among the scientists who pushed hard against the modeling frontiers were Philip Seiden and Humberto Gerola of IBM's Watson Research Center in New York. The pair were fascinated by the fact that spiral galaxies seemed to retain their pinwheel shape even though centrifugal force should disperse the whorls in the course of just a few rotations. In 1979, Seiden and Gerola hypothesized that new stars were replacing the old ones, forming new galactic arms that replenished the galaxies' spiral structure.

Of the two scientists, Seiden was more familiar with certain specialized modeling techniques, so to test their notion he set about fashioning a computer model that incorporated the mechanics of stellar procreation. (As an aging star goes supernova, astronomers believe, it sends out energy and shock waves that compress the surrounding interstellar gas, causing it to coalesce into new stars.) Seiden's first task was to translate the frame-

ENERGY STOREHOUSE

Some of the most powerful phenomena in the universe are produced by interactions of magnetic fields and plasma, a gaslike state of matter composed of charged particles. So-called cosmic jets, for instance, are narrow plasma streams millions of light-years long that seem to be ejected along magnetic field lines by some stars and galaxies. Closer to home, the Sun gives off a constant stream of charged particles called the solar wind. When the wind encounters a planetary magnetic field, it compresses the field on the sunward side and stretches it out on the planet's night side into the magnetotail, a kind of billowing windsock *(below).*

Because the solar wind is itself magnetized, the

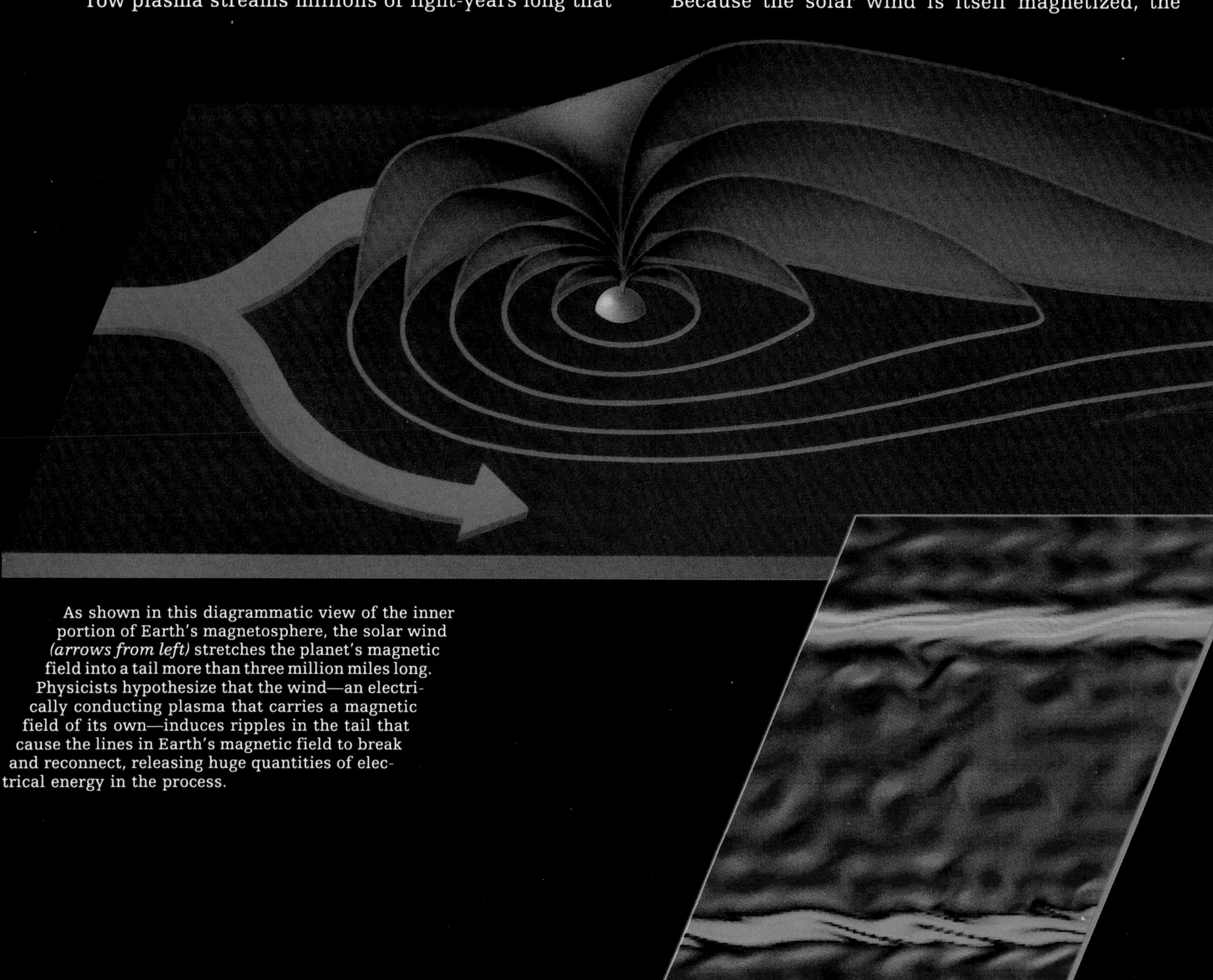

As shown in this diagrammatic view of the inner portion of Earth's magnetosphere, the solar wind *(arrows from left)* stretches the planet's magnetic field into a tail more than three million miles long. Physicists hypothesize that the wind—an electrically conducting plasma that carries a magnetic field of its own—induces ripples in the tail that cause the lines in Earth's magnetic field to break and reconnect, releasing huge quantities of electrical energy in the process.

In this first image from a computer simulation, the parallel edges of the magnetotail begin to twist slightly as the solar wind flows above and below the tail, raising small ripples, much as wind ruffles the surface of the sea.

meeting generates enormous quantities of electricity. When the energy stored in the magnetotail reaches overload, the tail releases it in a so-called magnetospheric substorm. During the substorm, the solar wind's magnetic field lines squeeze the magnetotail, pinching it in two and releasing a plasmoid—a huge structure of hot plasma held together by looping magnetic field lines—that is carried downstream with the solar wind. As the rest of the tail snaps back toward Earth, a blizzard of electrically charged particles slams into the upper atmosphere to create brilliant aurorae in the skies near both poles. Scientists wishing to understand more about these interactions in the Solar System and elsewhere simulate the process with fluid modeling programs *(pages 114-115)* on supercomputers. Some of the results are pictured at bottom.

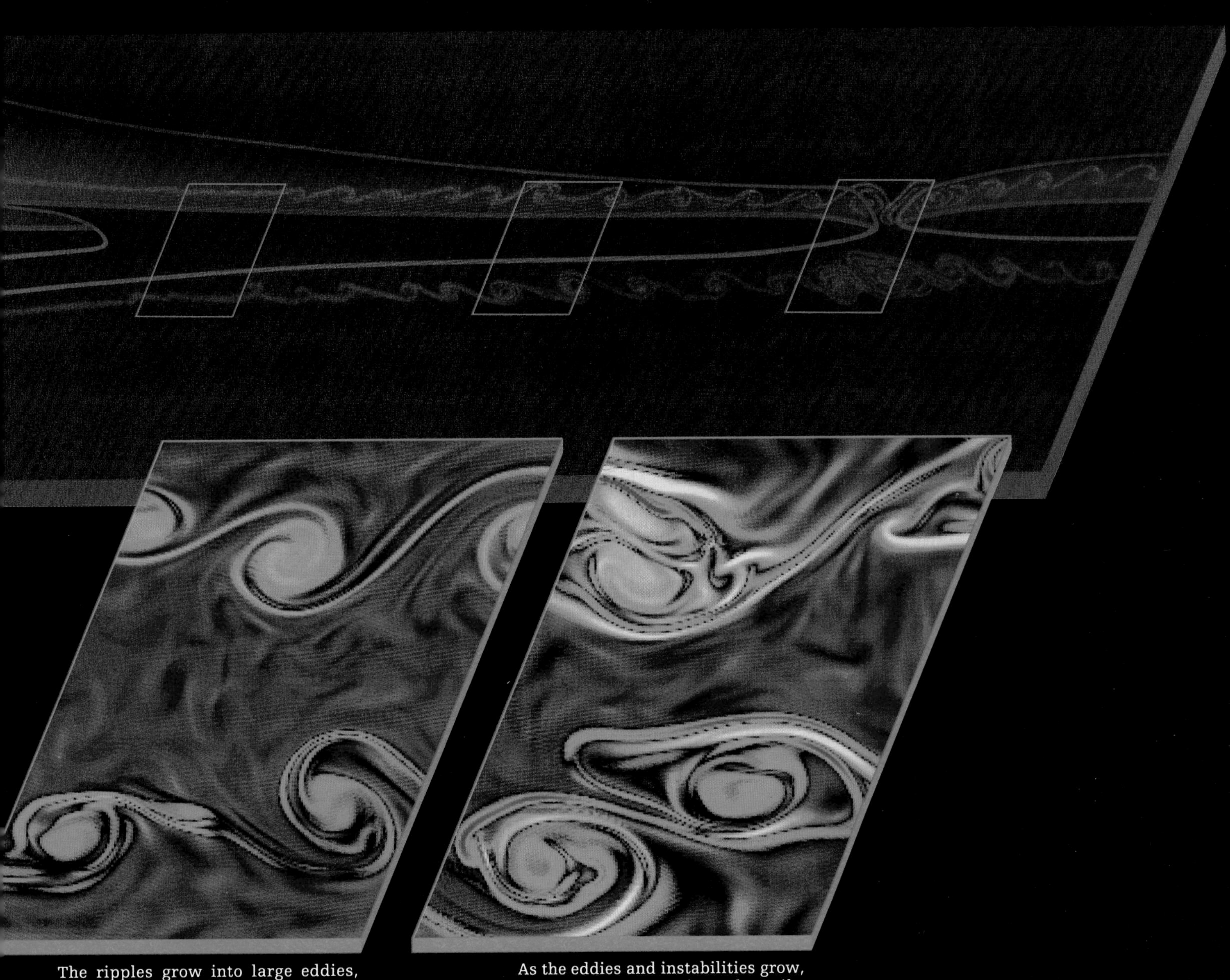

The ripples grow into large eddies, churning the edges of the magnetotail into an unstable region of fluctuating magnetic forces, electricity, and blobs of plasma.

As the eddies and instabilities grow, they extend deeper into the tail, stretching the magnetic field lines until they break. The resulting release of energy, predicted by the simulation is believed to have been corroborated by satellite observations of magnetospheric substorm activity.

work of a spiral galaxy into a form digestible by an IBM 3091 mainframe.

Knowing that most galaxies display differential rotation—that is, stars, dust, and gas in their inner portions revolve faster than matter near the perimeter—Seiden schematized his model galaxy as a set of fifty spinning concentric rings. He then programmed the inner rings to rotate faster than the outer rings. Every ring was in turn segmented into equal-size blocks, with each block demarcating a region of stellar activity. On his computer screen, a block that contained a filled circle, for example, signified a supernova about to trigger the formation of new stars in neighboring blocks.

Seiden's second task was to capture, in digital form, the propagation of star formation through a galaxy. He eventually settled on a complex statistical method, known as percolation theory, that geologists and electrical engineers have used to diagnose such flow problems as the seepage of liquids through porous rock and the migration of electrical signals through computer networks. The same theory, Seiden hoped, might describe the chain reaction of star births in galaxies and, ultimately, the seemingly impossible persistence of the spiral pattern.

After seeding his galaxy with a few random star clusters, Seiden set the model running. As the computer made its calculations over time, Seiden's program assessed the odds that a star in any block would turn supernova and create new stars in the blocks around it. The end product of the simulation was, after many hundreds of thousands of years, a circular map of stars with a neat, pinwheeling pattern. When Seiden stipulated an overall rotation rate for the model that matched the rate of a known galaxy, the result of the simulation closely mirrored the real thing.

THE SUPERCOMPUTER ASCENDANT

Though visually impressive, Seiden's model galaxies could only hint at the revolution in cosmic imaging to come. The primary instrument of the new realism—indeed, of a whole new generation of astronomical modelers—would be the supercomputer, an umbrella term for any machine whose design and performance represent a quantum advance over the technology of existing computers. The supercomputer's powers of image processing alone have yielded portraits that rival (and sometimes surpass) the best photographs obtained through optical telescopes.

Although supercomputers had been produced in small numbers since the early days of transistors, they had grown increasingly difficult and expensive to assemble. So complex was the wiring in IBM's Stretch supercomputer, for example, that the machine was completed nearly a year behind schedule. In fact, by the early 1970s the two leading supercomputer manufacturers, IBM and Control Data Corporation, had decided to shift their attention back to conventional mainframes, believing that demand for souped-up computers—and the profits to be expected from their sales—would always remain slim.

Almost alone in keeping the faith was Seymour Cray, an iconoclastic engineer who had been the chief designer of Control Data's supercomputer line. Since 1962, Cray had directed CDC's research-and-development efforts from a lab built especially for him near his home in rural Wisconsin. From this workshop overlooking the Chippewa River had issued the CDC 6600 and 7600 supercomputers, machines that secured Control Data's position in the marketplace. As CDC curtailed its research into supercomputers in 1972, Cray formed his own firm with the aim of capitalizing, as he put it, on "all the mistakes that the pioneers made."

Four years later, Cray lived up to his boast, unveiling a sleek, cylindrical, $8.8 million machine that avoided the design errors of earlier computers without introducing any new ones. Both the memory capacity and the processing speed of the Cray-1, as the new machine was called, were four times greater than those of the CDC 7600. For astronomers, such improvements betokened enormous savings in the run time of their models.

One beneficiary of the accelerated pace was Swiss astronomer Willy Benz. Whereas Benz's predecessors in computer modeling had used simulations to voyage far out into the universe, Benz set his sights on a target much closer to home: the Moon.

As a graduate student at the University of Geneva in the early 1980s, Benz had simulated the formation of stars from collapsing gas clouds on a UNIVAC 1161 mainframe, a descendant of the first commercial electronic computer. Although his models calculated the gravitational forces, magnetic forces, and pressure gradients of only 500 discrete points in a gas cloud, the UNIVAC required three or four weeks to run each simulation. Thus when Benz completed his degree and took a research post at the Los Alamos National Laboratory in New Mexico in the fall of 1984, his modeling ambitions swelled in proportion to the increased processing power that suddenly became his. Because he was a foreign researcher, Benz was not allowed time on machines designated for classified work, but he did have ample access to the so-called open Crays, which, as he noted later, "is more than you can get anywhere else."

AFTERMATH OF COLLISION

At Los Alamos, he became intrigued by a recently developed theory that the Moon was formed in the wake of a collision between Earth and another planet-size object. He therefore dreamed up an intricate computer model in which two large bodies were made to crash into each other at the high speed of eleven kilometers per second. So computationally intensive was the program that it ate up ten to twenty hours of processing time even on a Cray, but the effort was worth it. Benz's program produced some stunning, computer-generated moving images of the impact between a projectile—consisting of an iron core blanketed by a granite mantle—and a larger target of similar composition that was meant to represent the primordial Earth.

As Benz fiddled with the mass and makeup of the bodies in his model, he came up with a variety of different scenarios. The most arresting images by far were those produced when he posited a projectile having one-seventh the mass of the proto-Earth. In this case, the incoming object struck the Earth a glancing blow and then splintered, throwing debris into space. As the projectile's iron core was swallowed by the proto-Earth, the shattered material from its rocky mantle gradually accreted into a satellite with nearly the same mass and orbit as our present-day Moon. In supporting the new theory of lunar genesis, Benz's model contributed strong theoretical evidence against two competing explanations: that the Moon was spun off from Earth early in its history, and that the Moon was captured by Earth after sailing too near the planet on a journey from somewhere else in the Solar System.

THE COSMOS IN FREEZE FRAME

Benz's model had graphic significance as well. Thanks to the high-resolution moviemaking facility available on the Cray supercomputers at Los Alamos, the model yielded a time-lapse color movie of an episode of interplanetary violence. The movie could be examined in frame-by-frame stop action, a feature that enabled viewers to spot details that might have eluded them in a continuous animation.

Exercises like these whetted astrophysical appetites for pursuing such exotic cosmic quarry as quasars and galactic jets. Even the events governed by Einstein's daunting field equations, which describe the gravitational distortions of four-dimensional space-time by matter and energy, now seemed within the practical reach of computerized reenactment.

In the late 1980s, the search for more sophisticated graphics and the desire to simulate more exotic systems dovetailed in a computer model of epic scope. The architects of the model were Cornell University astronomers Saul Teukolsky and Stuart Shapiro, whose special interest was astronomical events

Cornell astrophysicists Stuart Shapiro *(left)* and Saul Teukolsky study images from their model of the birth of a black hole at the center of a star cluster. More than 10,000 pictures went into an eight-minute movie of the simulation.

involving what Teukolsky termed "extreme conditions": strong gravitational fields, high energies, and high velocities.

In 1986, the pair began probing the puzzle of quasars, the most energetic objects in the universe. Billions of light-years from Earth, they shine brighter than galaxies with hundreds of billions of stars, yet each is barely the size of our Solar System. The abiding mystery is how quasars form. Scientists suspect that at the core of every quasar there lurks a supermassive black hole, sucking up matter with a gravitational field so strong that not even light can escape it. The birth of quasars is therefore very likely tied to the birth of black holes.

Then, as now, a number of astronomers believed that black holes result from the collapse of unstable star clusters. Of course, direct observations of such an event were out of the question. "The collapse is such a short-lived phenomenon," Teukolsky and Shapiro wrote in 1988, "that catching a black hole in the act of formation is unlikely." To overcome this statistical impediment, they launched a numerical investigation into the kind of star-cluster collapse that might lead to black holes and quasars.

The computer they chose to run their model on was an IBM 3090-600 mainframe housed in Cornell's National Supercomputer Facility. Attached to the machine was a battery of array processors, special-purpose devices that speed a program's execution by tackling the most math-intensive calculations. With this digital arsenal at their disposal, Teukolsky and Shapiro performed 100 billion calculations describing the positions and velocities of 10,000 stars at 1,000 separate increments of time.

Soon afterward, the two researchers attended a scientific conference at the Drexel Institute of Technology in Philadelphia, where they screened, in effect, a home movie of their simulation—a rough sequence of scenes produced by the computer. In the audience was Craig Upson, a representative of a Hollywood computer-animation studio that had been founded by physicists and was known for creating special effects for such science-fiction movies as *2010* and *The Last Starfighter.* The studio, Digital Productions, owned what every astrophysicist lusted after: a Cray X-MP, the fastest supercomputer then in existence. "It was rather a curiosity," Shapiro later recalled, "that scientists couldn't get their hands on these things but cartoon illustrators had no trouble affording them." When the National Science Foundation offered to buy Teukolsky and Shapiro seventy-five hours of processing time on Digital's X-MP, the two astronomers packed their bags for Hollywood.

Upon arrival, Teukolsky and Shapiro toured Universal Studios, where Digital's simulation of Jupiter's swirling Great Red Spot for the movie *2010* was on display. To achieve the highest possible degree of realism, Digital's physicists and animators had based their model of the spot on the actual hydrodynamic equations that described the planet's surface currents. As Shapiro noted later, "It was a rather scientific means of generating cartoon images, and it impressed us."

The astronomers then teamed with Upson and graphic specialist Stefen

Fangmeier for four days to pound out a script for an animated simulation. Shapiro found the scripting process an odd way to deal with such weighty stuff as Einstein's equations and serious computations, but, as he acknowledged, "that's the only way to do a movie." Over the same period, the astronomers loaded the program for their model into the Cray X-MP, then returned to Cornell.

During the next few months, the pair regularly received thirty-five-millimeter rushes, or first film prints, from Digital, and screened them at a campus theater. When completed in 1986, the film ran just eight minutes, but a stupendous quantity of computations had made that segment possible. One scene alone, for example, entailed the motions of 7,000 particles in an area covered by 350 grid points over 4,000 increments of time. Modeling the velocities and positions of these particles required ten hours' labor by an array processor; converting the computations into moving pictures tied up the X-MP for another ten.

In July of 1988, Teukolsky and Shapiro summarized the events depicted by their animated model: "100 million neutron stars swirl around the center at velocities near the speed of light. Suddenly the delicate balance between their orbital motion and the inward pull of their mutual gravitational attraction is upset. Stars begin to rush toward the center. An avalanche ensues. Out of this catastrophic collapse a supermassive black hole arises. Gaseous debris accretes onto the black hole, radiating profusely before being swallowed. A quasar is born."

GALACTIC GEYSERS AND BEYOND

Although Teukolsky and Shapiro's model vividly dramatized the benefits of running astronomical models on supercomputers, most researchers still lacked regular access to the machines. Thanks to astrophysicists Michael Norman and Larry Smarr, however, that exclusiveness would soon dissolve.

Norman and Smarr had met in 1976 at the Lawrence Livermore Laboratory, where Norman was working under the tutelage of modeling pioneer Jim Wilson and Smarr was completing a model of colliding black holes. A few years later, during a visit to England's Cambridge University, Smarr came in contact with Martin Rees, one of the first astronomers to propose a theory for the origin of the strange phenomena known as galactic jets. Spurting at high speed from the center of some galaxies and quasars, these elongated streams of matter are among the largest coherent structures in the universe. They are thousands to millions of light-years in length and may be driven by the gravitational potential released by black holes. When Smarr returned to California in December of 1979, he persuaded Norman that the jets would make ideal candidates for modeling.

Norman opted to focus first on how galactic jets are created. Working on one of the lab's CDC 7600 supercomputers, he devised a simple model in which a jet's path of travel was represented on a square grid of 1,600 zones. Hydrodynamic equations tracked the gas ejected by a black hole at the center of the grid. Although this early simulation failed to re-create jets like those that had been observed, Norman soon got a chance to attack the problem again. Upon joining the Max Planck Institute for Physics and Astrophysics in Munich in 1980, Norman was given free range of the first Cray-1 installed in Europe. Smarr came to visit in June 1981, and the two joined with German researcher Karl-Heinz Winkler to launch a major modeling assault on the formation of galactic jets.

Smarr's arrival coincided with a Catholic holiday, in observance of which the computer center had closed for a week. But Winkler struck a deal with the center's computer operators: The technicians would receive twenty half-liters of beer, and in return, the modelers would be allowed unrestricted access to the institute's Cray-1. With Smarr providing tips on the behavior of matter under relativistic conditions, Norman ironed out the kinks in the model's software. This inaugural modeling session turned into a marathon; Smarr and Winkler eventually faded and left, but Norman pressed on, ultimately logging thirty-two continuous hours at the terminal.

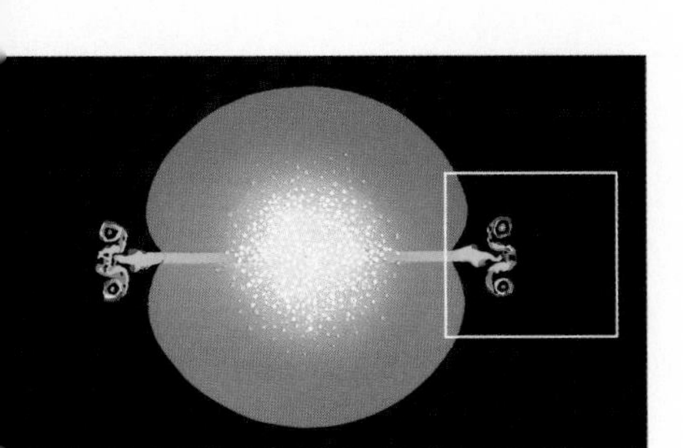

nodel developed by a team led by Michael rman of the National Center for Super- mputing Applications helps explain radio s that shoot from the nuclei of some ac- e galaxies. The high-energy particles start t as tightly focused beams, then spread o vast, chaotic lobes. According to the del, the beams begin to dissolve as they ss a shock wave formed when a spherical ell of gas and dust emanating from the axy is compressed as it collides with mat- in intergalactic space *(above).* The mputer-generated images below depict the reasing turbulence and disintegration of a as it pierces the shock front.

Twenty-four hours into the session, Norman's lack of progress with the model prompted him to scrap the 1,600-zone grid he had conceived at Livermore and start afresh. With a boldness born of exhaustion, he designed a two-dimensional rectangular grid containing 50,000 zones. The gamble paid off in a series of computer-generated images showing galactic jets in all their swirling, interstellar glory *(below).* Once the three researchers had caught up on their sleep, they festooned the walls of the institute with the pictures that had resulted from the model run.

A year later, Smarr returned to the institute for a workshop on astrophysical modeling, which Norman and Winkler had organized under the auspices of NATO, the North Atlantic Treaty Organization. During the workshop, the German sales representative for Cray Research touted the company's latest model, the X-MP, and its four-million-word memory—large enough to model galactic jets in three rather than two dimensions. Sitting in a beer hall outside Munich after the presentation, Winkler challenged Smarr—

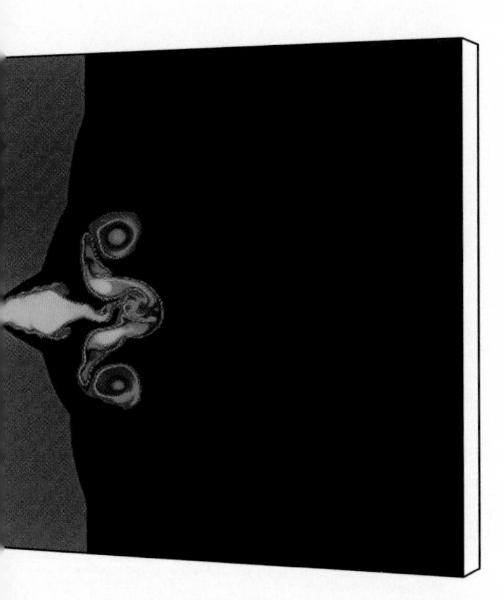

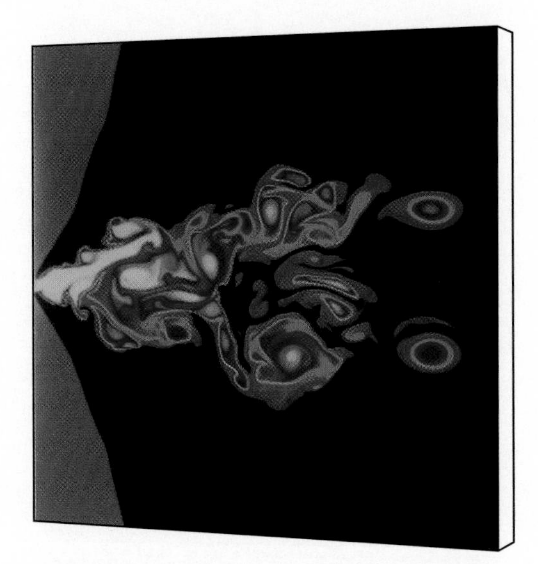

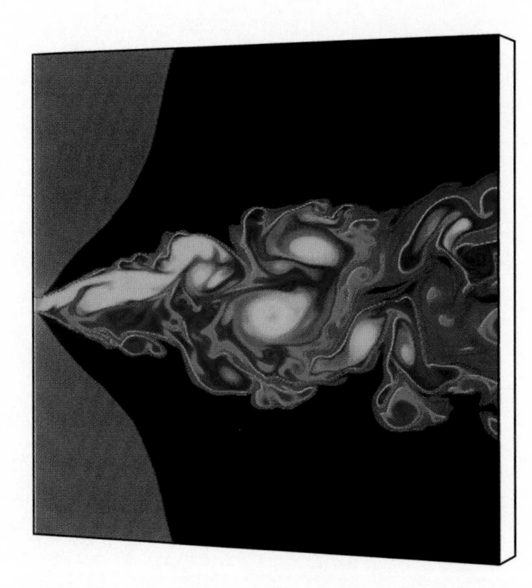

now a professor at the University of Illinois—to "get an X-MP to Illinois."

Stateside, Smarr wrote a manifesto entitled "The Supercomputer Famine in American Universities." In it, he warned that Germany and Japan were beating the United States in the supercomputer race, possibly imperiling national security. But Smarr did not stop there. In 1983, he presented the National Science Foundation with an unsolicited proposal, which became known as the Illinois Black Proposal for the color of its cover. The document, written with Norman and Winkler, requested a grant of $43.5 million to establish a supercomputer facility at the University of Illinois. To illustrate just how far the country had fallen behind, Smarr also circulated some color movies of galactic jets that Winkler had recently produced on the Max Planck Institute's X-MP. Shortly thereafter, the foundation announced its intention to open five national supercomputer centers—one of them to be located on the university campus in Urbana-Champaign.

With supercomputers now in easy reach through the National Supercomputing Centers, as the new facilities are called, astrophysicists aspire to even more advanced models. Simulating the explosion of a supernova in three dimensions, for example, promises to reveal processes that astronomers cannot observe in the course of a single lifetime. As supercomputers and other highly sophisticated instruments beckon scientists to embark upon ever-lengthening imaginary journeys through the cosmos, it is clear that the study of the stars has progressed a long way since the days of Erik Holmberg and his laboratory lamps. The confines of deep space—those once-forbidden celestial precincts that astronomer Mike Norman has called "the unreachable points in the universe"—may finally be within reach.

SIMULATING COSMIC EVENTS

The universe is, above all, a process—a continuously unfolding tale of complex interactions and changing conditions. For the most part, however, observing instruments can capture only brief incidents in a lengthy history or the mere aftereffects of events obscured by distance or time. Among the many sophisticated tools of the new astronomy, the computer alone is capable of putting the universe in motion by manipulating time. Through carefully programmed simulations, astrophysicists can animate their speculations on how things got to be the way they are, watching millions or even billions of years of cosmic evolution play out on a computer screen in a matter of minutes.

The simulation process begins with the fashioning of an intricate mathematical model of the celestial objects and forces under study *(pages 110-115).* By repeatedly solving the model's equations, the computer alters the arrangement of objects in steps, moving the simulation forward at a rate designed to reveal significant change. Complex graphics-generating techniques render the results as a sequence of vivid images much easier to comprehend than the numbers and formulas they represent. To test validity, astronomers compare these images with actual observations: The closer the match, the better the chances that the simulation has painted an accurate picture.

Simulations offer insights on a wide variety of cosmic phenomena. The examples on the following pages include depictions of coalescing galaxies and the formation of a supermassive black hole, as well as a possible solution to the longstanding riddle of the Moon's origin.

GALACTIC METAMORPHOSIS

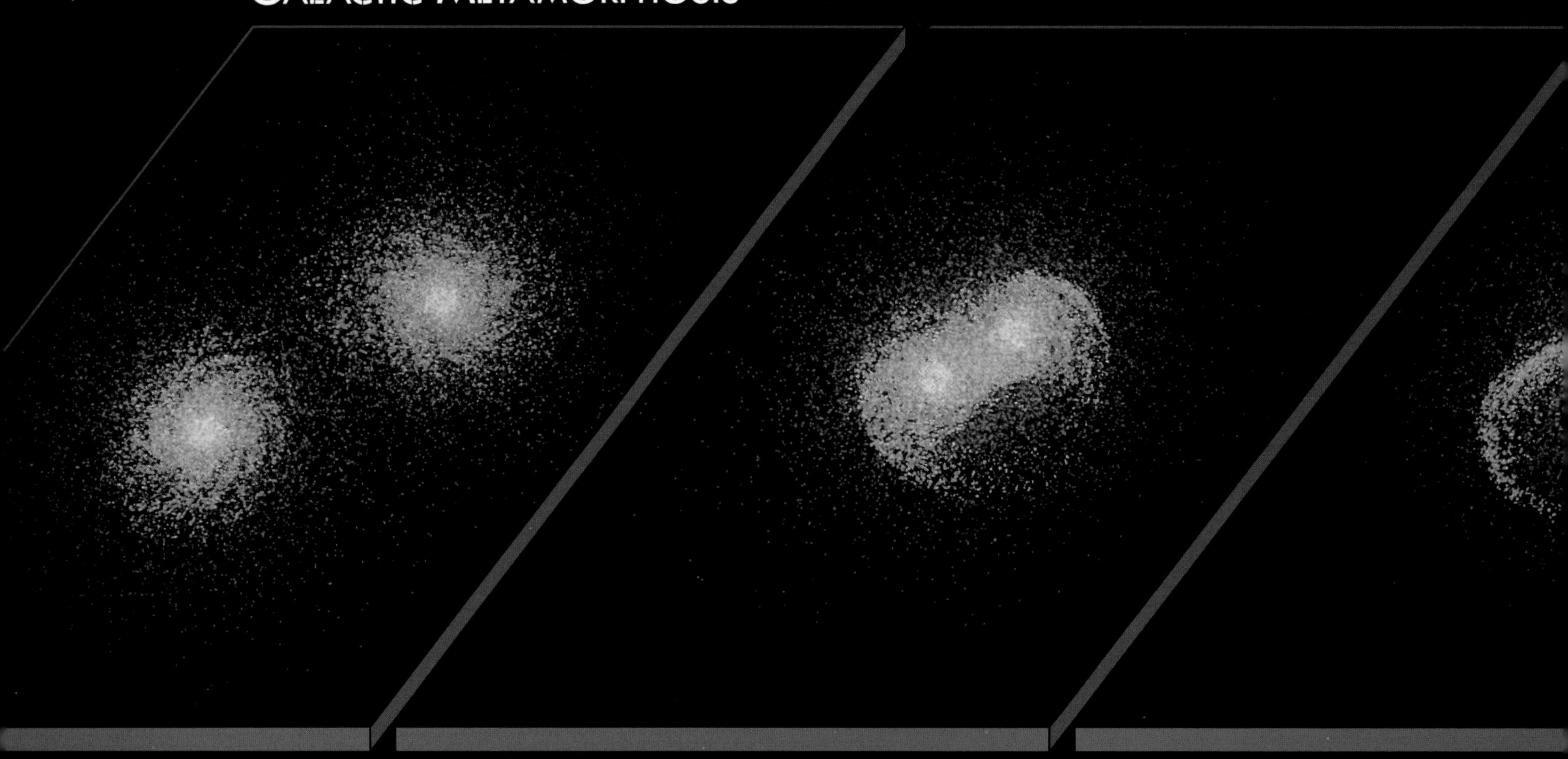

In a sequence spanning 560 million years, two disk galaxies merge and begin to separate, sending out long streamers of stars that closely resemble actual o

SIX INTO ONE

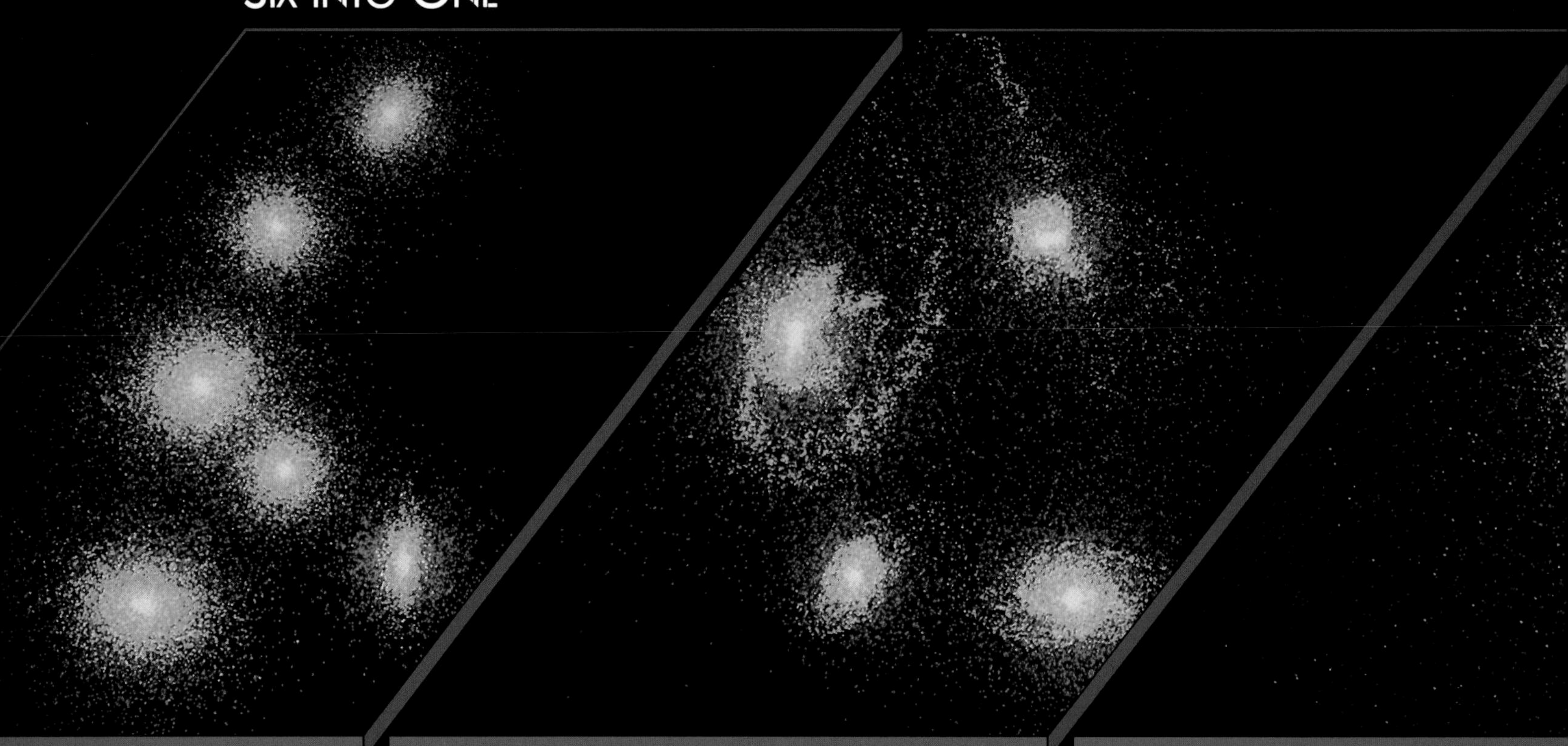

Thousands of individually modeled particles produced this simulation of six disk galaxies coming together over the course of nearly three and a half billion yea

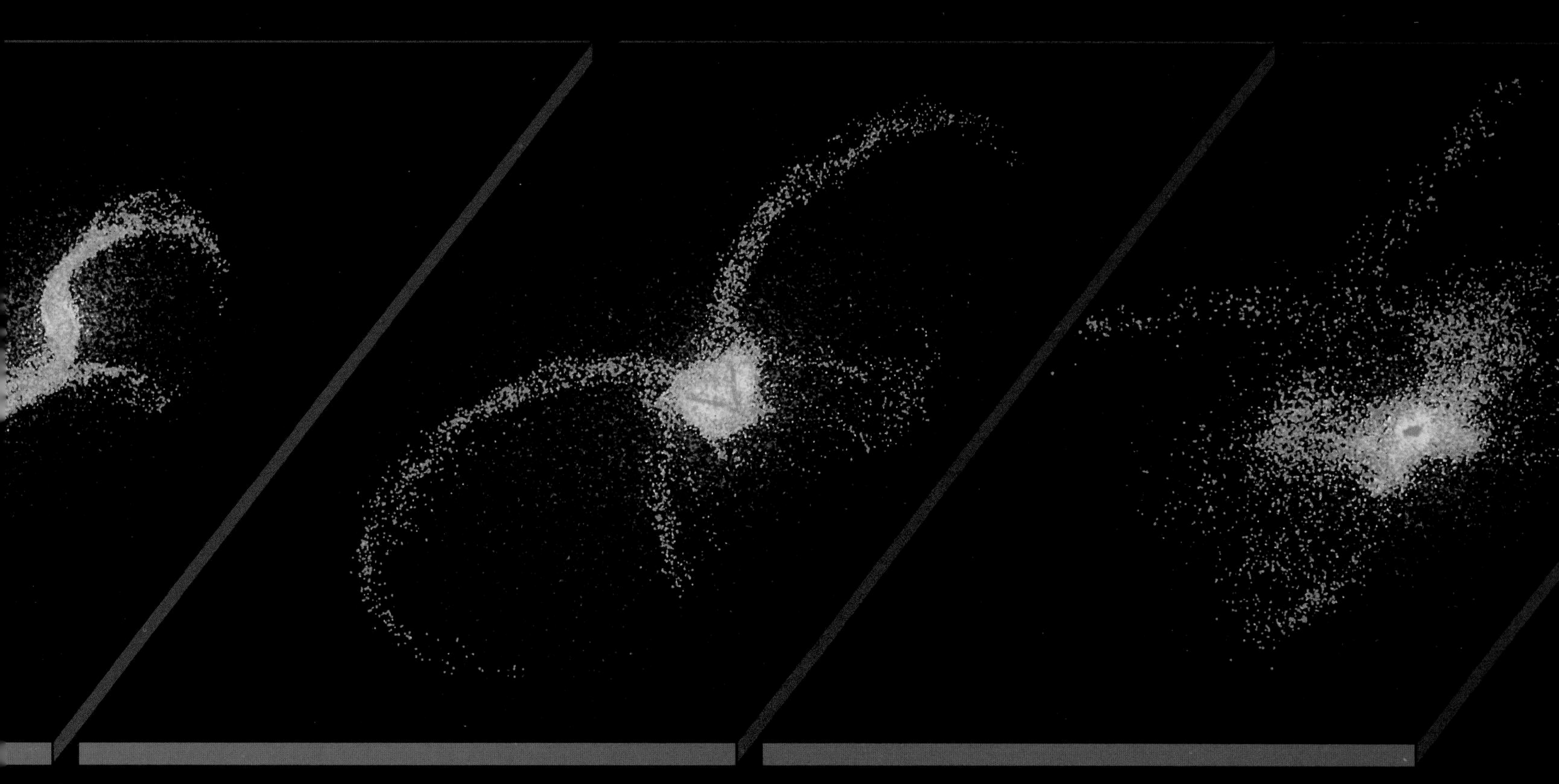

ervations. Red in the images represents invisible dark matter that contributes to the strange gravitational interactions.

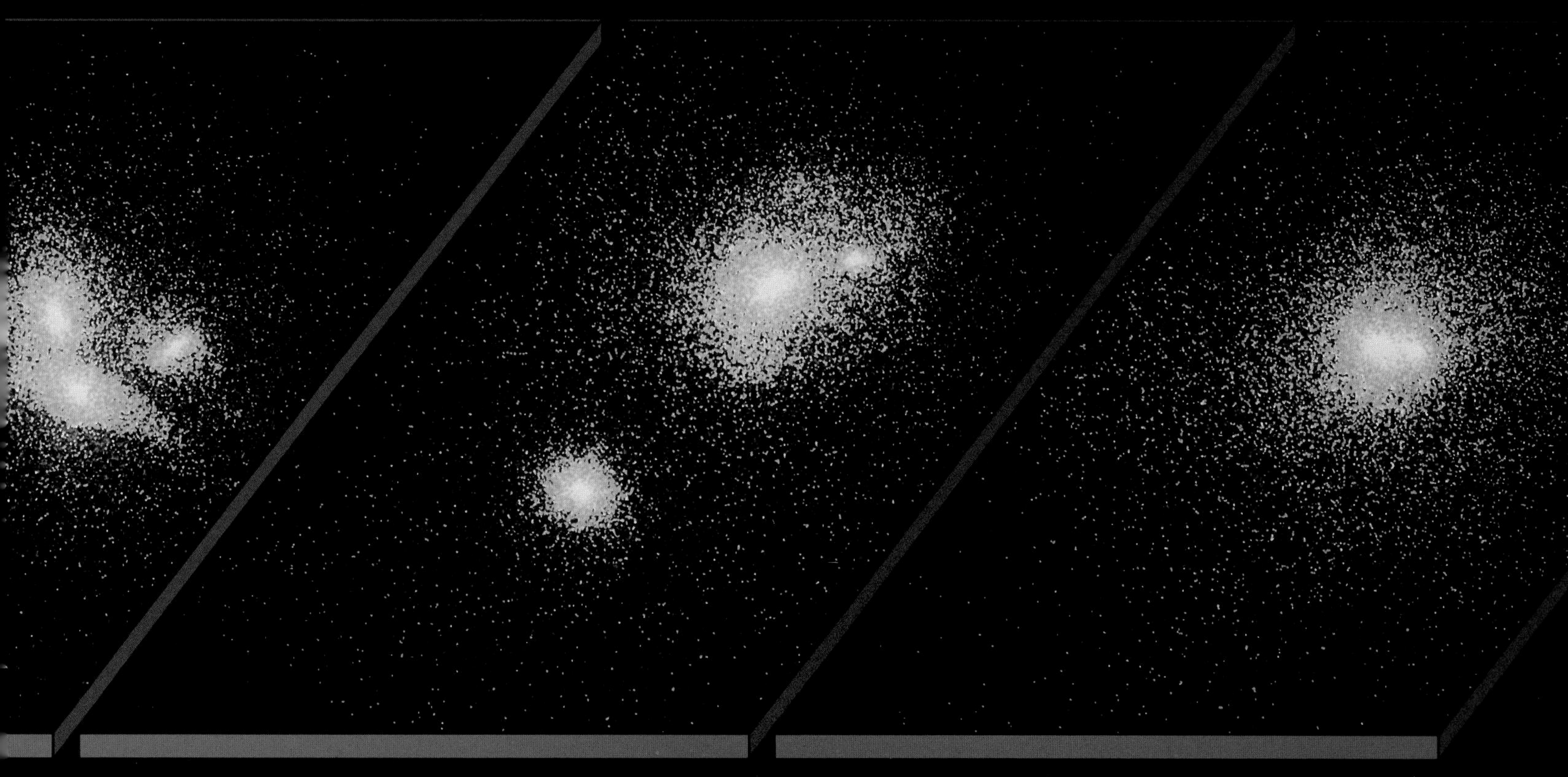

uch compact groupings are rare but hold important clues as to how other unusual types of galaxies are formed.

Re-Creating the Moon

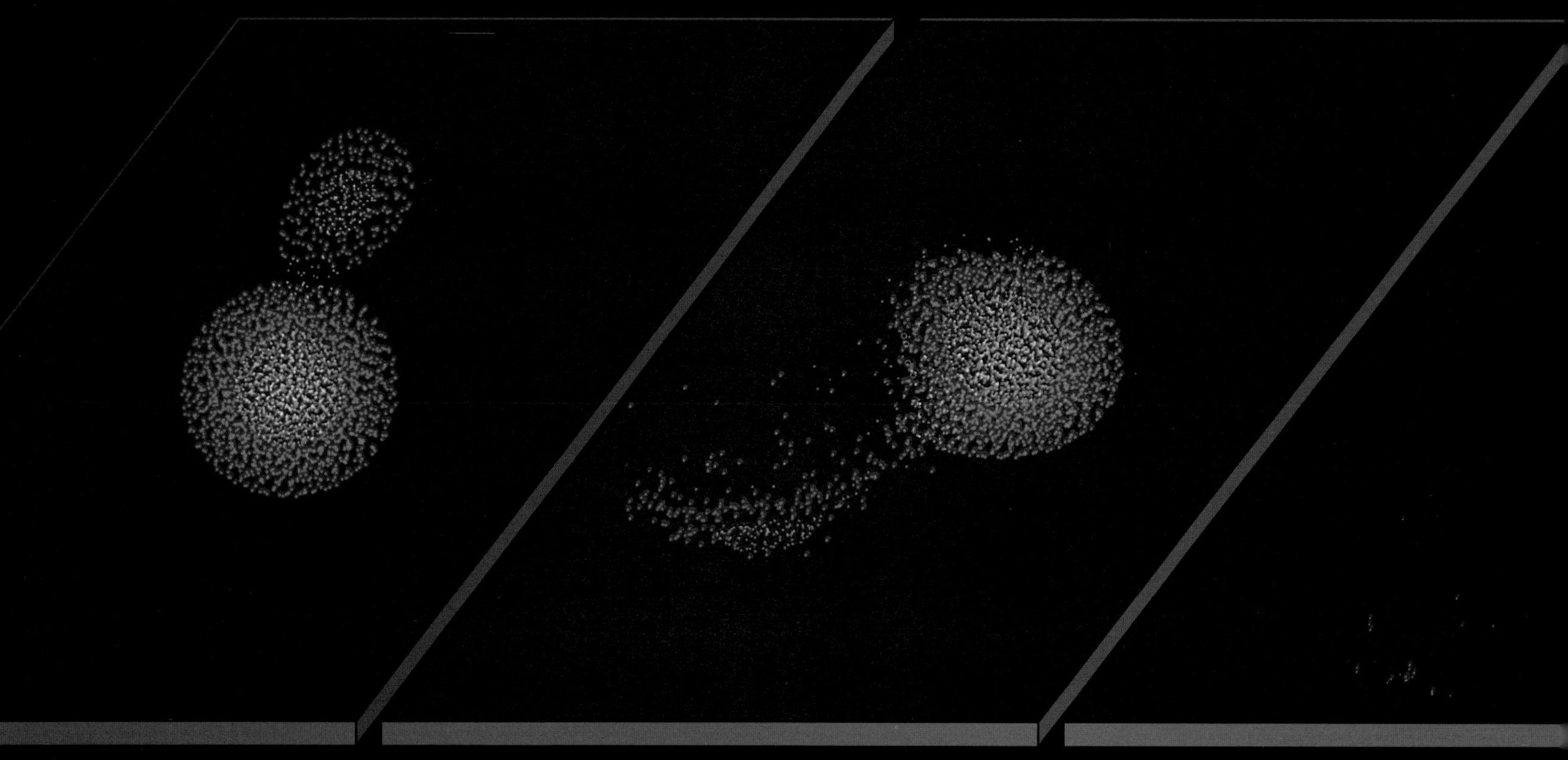

In a scenario that may account for the Moon's lack of iron, two bodies with iron cores *(pink)* and granite mantles *(red)* collide. Long-range views in the last thre

Birth of a Black Hole

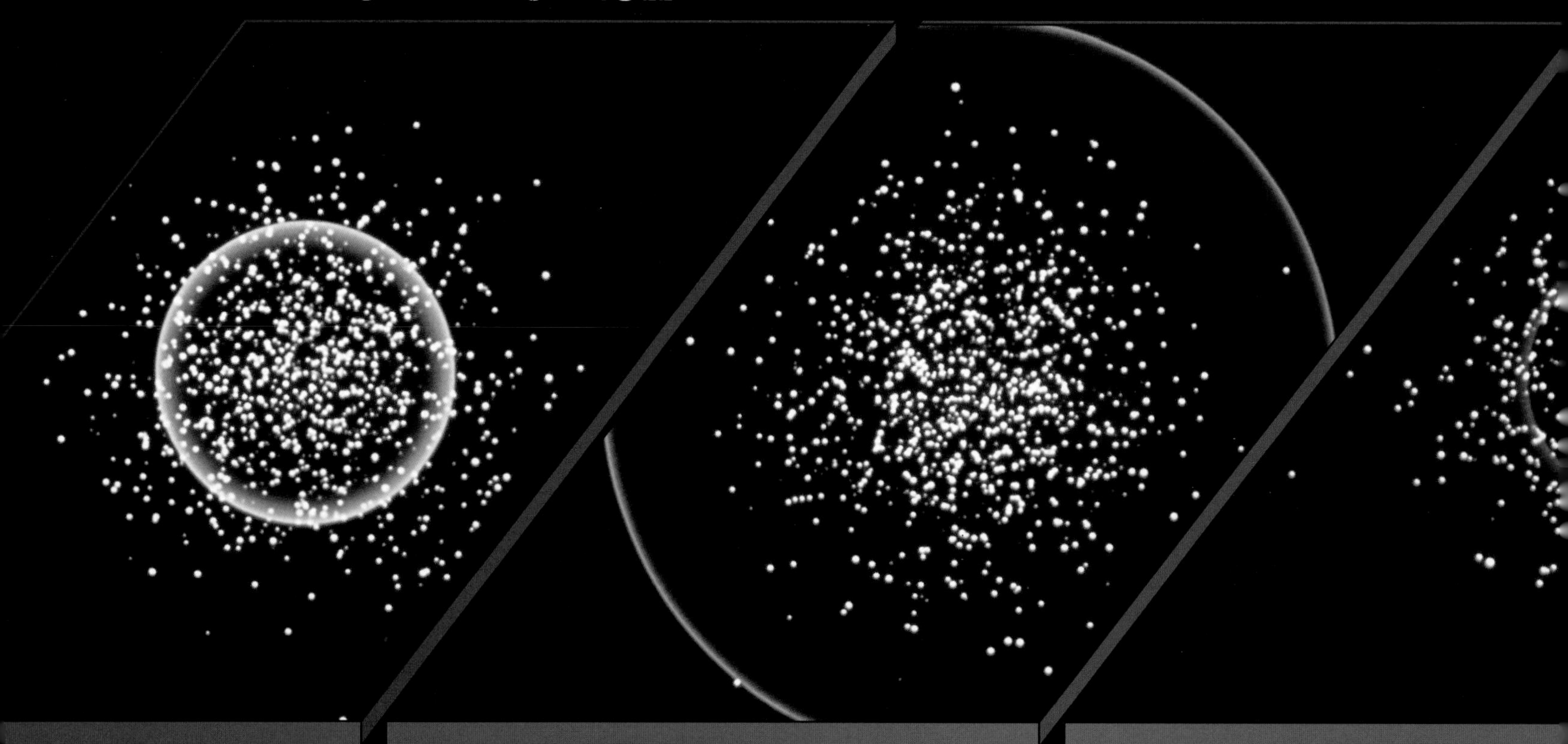

An unstable cluster of neutron stars begins to collapse, forming a black hole *(small central circle)* that grows as it sucks in more matter. Concentric rings in th

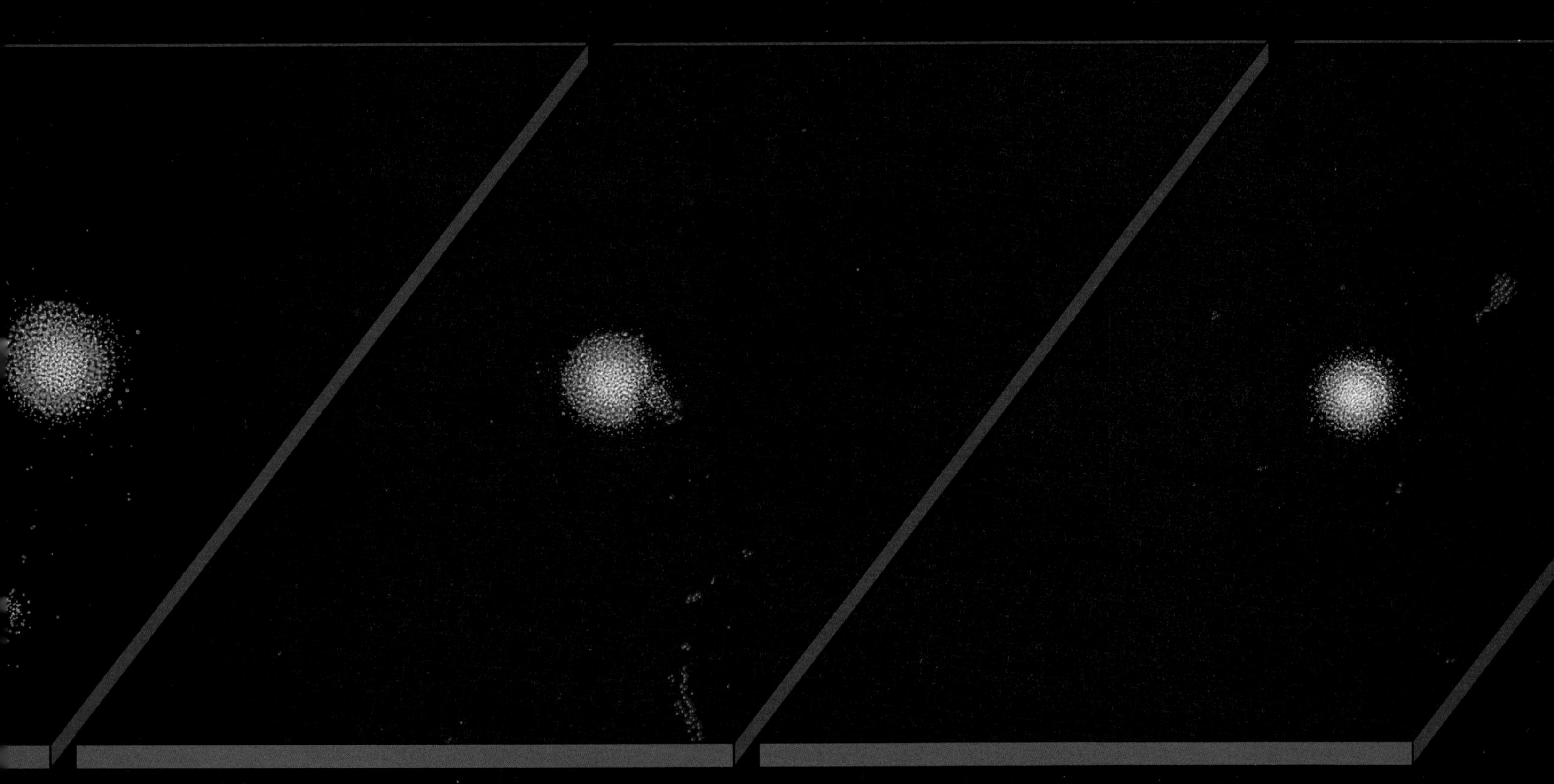

ames show the smaller core merging with the larger proto-Earth, while its stripped-off mantle remains in orbit to form a satellite.

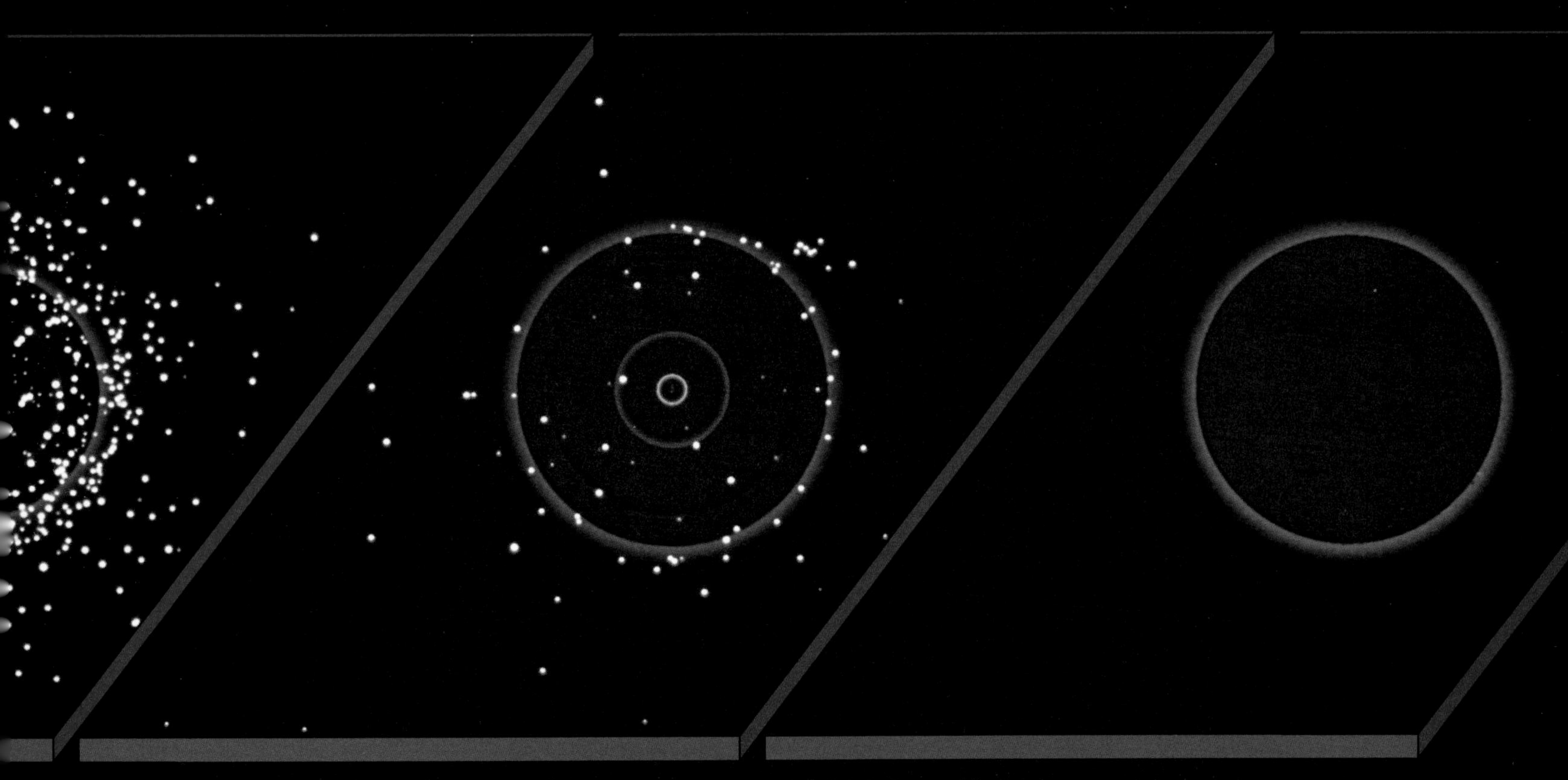

nages represent light that is at first able to escape but is then trapped by the black hole's inexorable gravity.

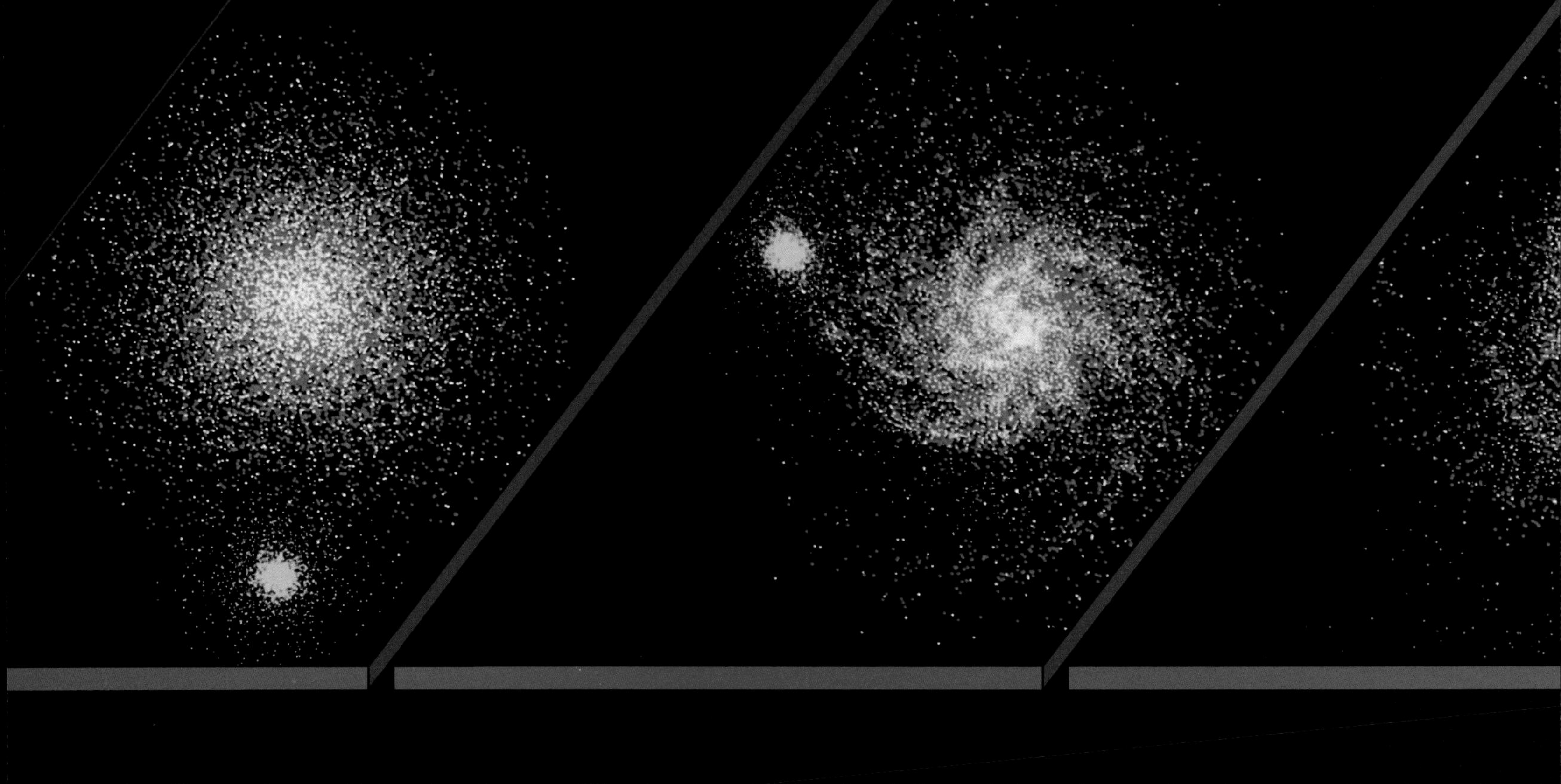

The series of images above and below show from two different angles a satellite galaxy *(yellow)* orbiting and eventually being consumed by a larger disk galax

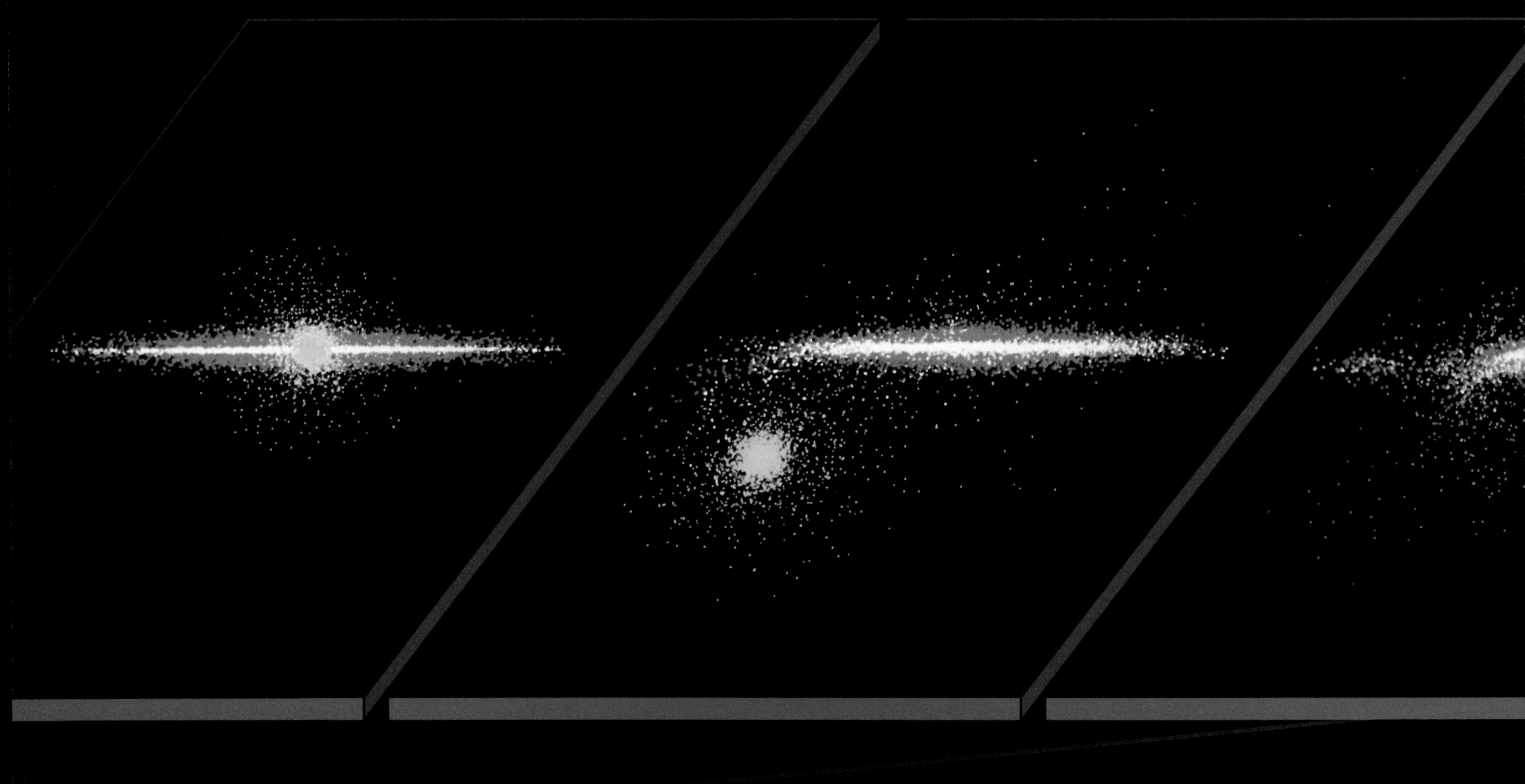

ue and white). The edge-on views below reveal how the satellite's tilted orbit affects the large galaxy's orientation.

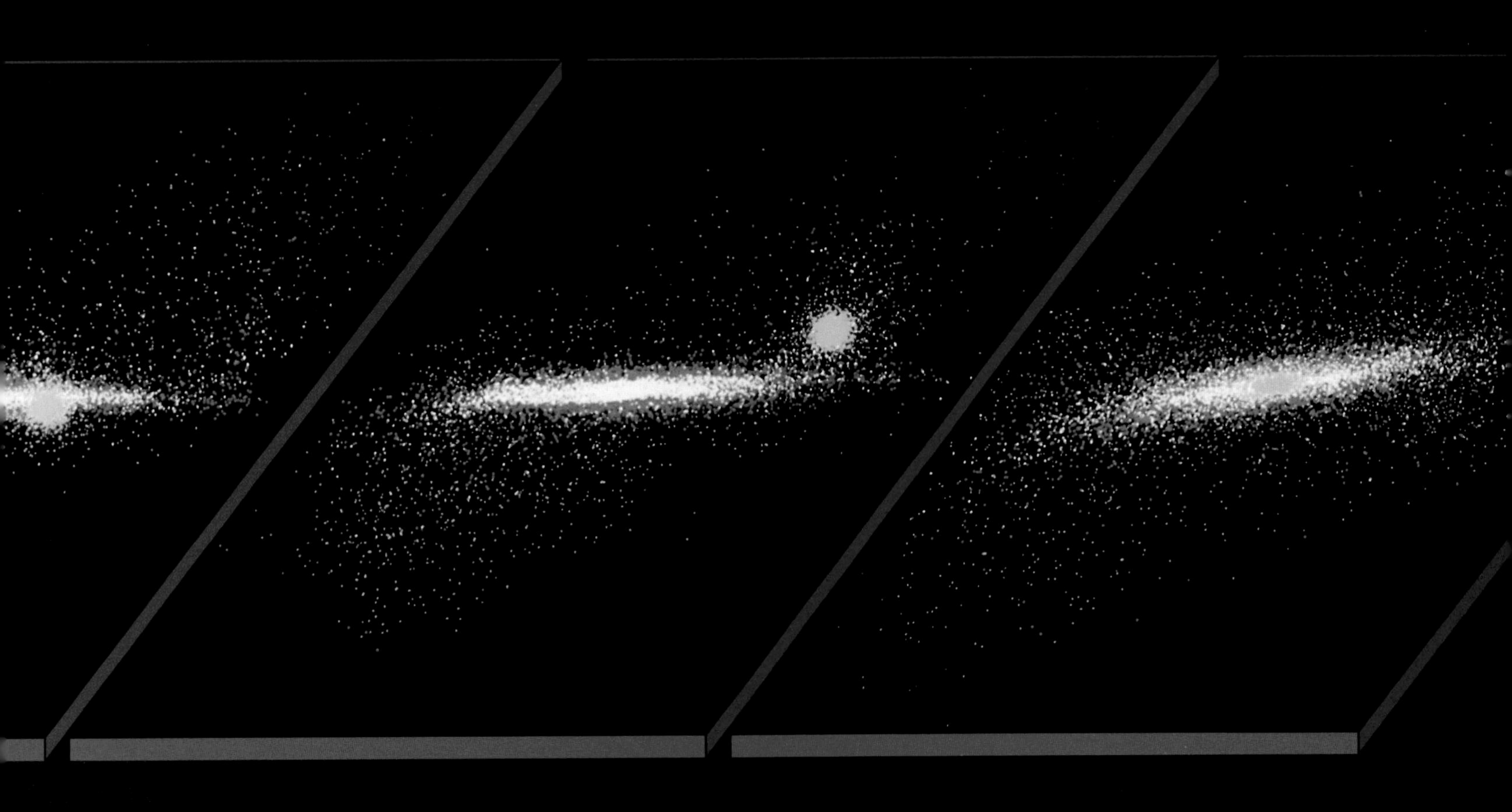

GLOSSARY

Absorption line: a dark line or band at a particular wavelength on a spectrum, formed when a substance between a radiating source and an observer absorbs electromagnetic radiation of that wavelength. Different substances produce characteristic patterns of absorption lines.
Accretion disk: a disk formed from gases and other materials drawn in by a compact body, such as a black hole or neutron star, at the disk's center.
Aperture synthesis telescope: a radio telescope that uses several antennas, which are moved between repeated observations of the same celestial object, to simulate a single observation made by one much larger antenna.
Argon: an inert, gaseous element, used in electronic instruments that detect high-energy radiation such as x-rays and gamma rays.
Astronomical Image Processing System (AIPS): computer software used for compiling and processing images, especially those from an array of two or more telescopes.
Beryllium window counter: an x-ray detector made up of several proportional counters beneath a thin foil of the metal beryllium, which blocks other types of radiation.
Binary accretion: a process in which one star in a binary system draws matter from its companion by gravitational attraction.
Binary system: a gravitationally bound pair of stars in orbit around their mutual center of mass. Binary stars are extremely common, as are systems of three or more stars.
Bipolar flow: the outward movement of gas from a young star in two opposed streams.
Blackbody: an object that absorbs all radiation reaching it and also emits radiation, with the brightest wavelength depending only on the temperature of the object itself.
Black hole: theoretically, an extremely compact body with such great gravitational force that no radiation can escape from it. Proposed varieties include mini black holes, low-mass objects from the beginning of the universe; stellar black holes, which form from the cores of old, very massive stars; and supermassive black holes, equivalent to several hundred million stars in mass and located in the centers of galaxies.
Bolometer: an instrument used to measure radiant energy, especially microwave and infrared radiation.
Carbon monoxide: a gaseous compound whose molecules each contain one carbon atom and one oxygen atom. Easily detected in space, carbon monoxide gives astronomers a means of mapping interstellar clouds that also include dust and other gases.
Centrifugal force: the apparent outward force felt by a body rotating about an axis.
Cerenkov radiation: radiation emitted by a particle moving through a medium faster than the speed of light in that medium. High-energy cosmic rays produce Cerenkov radiation, detectable as faint blue light, when they strike Earth's atmosphere.
Charge-coupled device (CCD): an electronic array of detectors, usually positioned at a telescope's focus, for registering electromagnetic radiation.
Collimator: one or more lenses or mirrors used inside a spectrograph or other instrument to make the light focused by a telescope parallel; also, in x-ray instruments, a device that restricts the x-rays admitted to a detector to those coming from a single direction in space.
Cosmic ray: an atomic nucleus or other charged particle moving at close to the speed of light; thought to originate in supernovae and other violent celestial phenomena.
Cyclotron radiation: radiation emitted as electrons accelerate in spiral paths within a magnetic field.
Dark matter: a form of matter that has not yet been directly observed but whose existence is deduced from its gravitational effects.
Electromagnetic radiation: waves of electrical and magnetic energy that travel through space at the speed of light.
Electromagnetism: the phenomena associated with electrically charged particles in motion, including electric and magnetic fields and the forces that they exert. Electromagnetism does not affect neutral particles such as neutrinos.
Electron: a negatively charged particle that normally orbits an atom's nucleus but may exist in isolation.
Emission line: a bright band at a particular wavelength on a spectrum, emitted directly by the source, and indicating by its wavelength a chemical constituent of that source.
Fluid modeling: a technique for simulating the behavior of a nonsolid celestial object, using a computer and principles of fluid dynamics to track matter across a grid of cells.
Free-free radiation: radiation emitted by a negatively charged electron when its free movement through space is slowed by its attraction to a positively charged ion.
Frequency: the number of oscillations per second of an electromagnetic or other wave. *See* Wavelength.
Fusion: the combining of two atomic nuclei to form a heavier nucleus, releasing energy as a by-product.
Galactic plane: the central plane of the Milky Way galaxy; also, the central plane of any disk-shaped galaxy.
Galaxy: a system that contains anywhere from millions to hundreds of billions of stars as well as varying quantities of gas and dust.
Galvanometer: an instrument that detects and measures small direct electric currents.
Gamma rays: the most energetic form of electromagnetic radiation, with the highest frequency and the shortest wavelength. Because the Earth's atmosphere absorbs most radiation at that end of the spectrum, most gamma ray astronomy is performed from space.
Geosynchronous: describing the orbit of a satellite that completes one revolution around the Earth in twenty-three hours and fifty-six minutes, the same time Earth takes to turn once on its axis; thus the satellite remains above one region on the ground.
Germanium: a chemical element commonly used in semiconductor detectors and spectrometers.
Gravitational lensing: an effect of general relativity in which the gravity of a very massive object bends the radiation of an object behind it, distorting its apparent image and often producing one or more duplicate images.
Gravity wave: a theoretical perturbation in an object's gravitational field that would travel at the speed of light. Relativity theory predicts that gravity waves may result from accelerating, oscillating, or violently disturbed masses, categories that include black holes, neutron stars, and supernovae.
Gravity wave antenna: an instrument designed to detect gravity waves, typically incorporating cryogenically cooled metal bars or laser interferometers that operate in a vacuum.
Grazing-incidence mirror: a telescope mirror designed to reflect x-rays and extremely short ultraviolet rays at very shallow angles and thus focus them. At steep angles, the

rays would be absorbed by the mirror materials and thus not be focused or detected.
Hubble's law: the astronomical formula stating that widely separated objects, such as distant galaxies, recede from one another at a rate proportional to the distance between them. The Hubble constant, based on the law, is an estimate of the rate at which the universe is expanding at the present time.
Hydrodynamics: the study of motion in fluids, meaning liquids or gases.
Hydrogen: the most common element in the universe. Stellar energy comes primarily from the fusion of hydrogen nuclei.
Hydroxyl radical: a molecule made up of one hydrogen atom and one oxygen atom. This interstellar molecule was the first to be identified by radio astronomy, in 1963.
Imaging proportional counter: a sophisticated instrument that registers x-radiation received by a telescope to produce a picture of the radiation source. Early proportional counters lacked this image-making capacity.
Infrared: a band of electromagnetic radiation with a lower frequency and a longer wavelength than visible red light. Most infrared radiation is absorbed by the Earth's atmosphere, but certain wavelengths can be detected from the ground.
Interferometry: the location and examination of sources of electromagnetic radiation through the simultaneous use of two or more separated telescopes. Interferometers produce overlapping wave patterns from the radiation; the patterns are studied to determine the brightness and angular structure (diameter, presence of jets, etc.) of the emitting source.
Inverse-square law: the mathematical relationship that describes the change in brightness of a star or other point source of light, which occurs inversely in proportion to the square of the distance from the source; also, any similar mathematical formula that describes how certain forces such as gravity change in strength with distance from a central point.
Ion: an atom that has lost or gained one or more electrons and has become electrically charged. In comparison, a neutral atom has an equal number of electrons and protons, giving the atom a zero net electrical charge.
Isotope: one of two or more forms of a chemical element that have the same number of protons but a different number of neutrons in the nucleus.
Jet: a narrow stream of high-speed matter ejected by a star, binary system, or galaxy. Many kinds of radiation, including radio waves and visible light, may be emitted by a jet or by interstellar or intergalactic matter that is disturbed by the jet.
Kelvin: the name give to an absolute temperature scale in which 0 is at absolute zero, about -273 degrees Celsius, and in which a unit of temperature, called a Kelvin, is equal to a Celsius degree.
Kinetic energy: an object's energy of motion.
Laser interferometer gravitational telescope: an apparatus designed to detect gravity waves, consisting of mirrors suspended in a vacuum whose relative positions are constantly monitored by laser beams. Slight changes in the mirror positions could indicate the passage of a gravity wave.
Magnetometer: a device for measuring the strength and direction of a magnetic field.
Magnetosphere: the region around a planet in which electrified particles are trapped in the planet's magnetic field.
Magnetotail: a billowing, streamlike extension of the magnetosphere of Earth or another planet, formed on the planet's dark side by the action of the solar wind.
Mainframe: the largest type of computer, usually capable of processing data for a variety of applications and users simultaneously.
Nanometer: a unit of measurement equal to one-billionth of a meter.
Nebula: a cloud of interstellar dust or gas, in some cases a supernova remnant or a shell ejected by a star.
Neutrino: a chargeless, subatomic particle with little or no mass.
Neutron star: a very dense body composed of tightly packed neutrons; one possible product of a supernova explosion. Neutron stars are observed as pulsars.
Nova: in ancient times, a very bright star that appeared where one was never seen before, and which soon faded away. In modern astronomy, a binary system with a white dwarf star as one of its members; eruptions occur that temporarily make the system thousands of times brighter than normal.
Nucleosynthesis: the creation of atomic nuclei; for example, by fusion of the nuclei of lighter atoms. In theory, heavier elements evolve from hydrogen as a consequence of high-energy particle collisions in the interiors of stars.
Ozone: a highly reactive, unstable molecule consisting of three atoms of oxygen.
Parabolic reflector: in astronomy, a reflector in a telescope that is shaped like a mathematical paraboloid and that reflects parallel rays of visible light or other radiation from a celestial object to a single focus.
Particle: the smallest component of any class of matter; for example, the elementary particles within an atom (such as electrons, protons, and neutrons), the smallest constituent of a gas (atoms and molecules), or the smallest forms of solid matter in space (interstellar and interplanetary dust particles).
Particle accelerator: a device, often several miles in length or circumference, used to accelerate subatomic particles to high velocities and fire them at target atoms and other particles. The results of collisions suggest the particles' properties.
Particle-mesh model: a computer model that studies the interactions of particles within a uniform three-dimensional grid structure superimposed on them.
Particle-tree model: a computer model that studies the interactions of particles within a hierarchical, three-dimensional structure made up of cells of different sizes.
Peculiar galaxy: a galaxy that differs from the standard types through the presence of unusual structures, such as jets or rings, or of unusual features in its spectrum.
Percolation theory: a statistical method used to study the characteristics of slow fluid movement through another medium; used in some computer simulations of star formation.
Photocell: a device that absorbs light and produces a measurable current, allowing the detection and calibration of light sources.
Photomultiplier tube: a device that detects very faint sources of visible light by amplifying the flow of electrons produced when the light strikes a photoelectric surface. For each incoming photon of light, a million or more electrons flow through the tube in a measurable current.

Photon: a unit of electromagnetic energy associated with a specific wavelength. It behaves as a chargeless particle traveling at the speed of light.
Pixel: short for picture element; one of the thousands of dots that make up a digital image. The intensity of radiation at each position in the image can be represented by assigning a false color (one that may not correspond to the real color of the object) to the pixel at that position.
Plasma: a gas of ionized particles, in contrast to ordinary gases, which are electrically neutral. Plasmas are sensitive to electrical and magnetic fields and are considered to be a fourth state of matter, along with ordinary gases, liquids, and solids.
Plasmoid: a moving object consisting of a plasma contained within a closed magnetic field structure; for example, a portion of Earth's magnetotail that separates from its surroundings and speeds away from Earth.
Polarization: the tendency of some electromagnetic waves to vibrate preferentially in a single plane rather than uniformly in all directions perpendicular to their motion. Polarization is produced by the source of the radiation and by the medium through which it travels.
Positron: the antiparticle of the electron, having the same mass as an electron but an equal and opposite (i.e., positive) electrical charge.
Proton: a positively charged particle, normally found in an atom's nucleus, with 1,836 times the mass of an electron.
Protostar: a large, gaseous sphere, held together by its own gravitational attraction, that shrinks and compresses to become a star.
Pulsar: a radiating source from which bursts of energy are received at precisely spaced intervals of several seconds or less. Pulsars are thought to be rapidly rotating neutron stars with very strong magnetic fields.
Quasar: shortened from quasi-stellar object; an extremely powerful, bright source of energy, located in a very small region at the center of a distant galaxy, that outshines the whole galaxy around it.
Radio galaxy: a galaxy that is an unusually strong source of radio waves and is distinguished from a radio-emitting quasar in that the source is not concentrated in a tiny region at the center.
Red giant: an aging star of relatively low mass whose atmosphere has greatly expanded and cooled, so that it appears orange or red.
Red shift: a Doppler effect seen when a radiating source recedes from the observer. The received waves lengthen so that any absorption or emission lines move from their expected frequencies. In visible light, this shift is toward the red end of the spectrum.
Reflector: mirror or other surface used to focus radiation; also, a telescope using such a focusing surface.
Resolution: the degree to which details in an image can be separated, or resolved. The resolving power of a telescope is usually proportional to the diameter of its mirror or aperture.
Scintillation counter: a device that determines the number, distribution, and source of gamma rays by passing them through a crystal structure.
Semiconductor: a substance with properties of electrical conductivity that change with changing conditions, such as temperature or the absorption of energy.
Sinusoidal wave: a regularly curving wave describing the behavior of the simplest kind of oscillating system. Pure musical tones or electrical signals produce sinusoidal waves.
Smooth-particle hydrodynamics: a method of analyzing motion in a fluid, such as a gas cloud, by modeling the trajectory of individual particles within the fluid mass.
Solar wind: a current of charged particles that streams outward from the Sun.
Space-time: a four-dimensional system in which any physical quantity can be located by specifying its position in three spatial dimensions plus time; also, the physical reality that exists within this system.
Spark chamber: a device that detects gamma rays and may also be equipped to track their trajectories.
Spectral energy curve: *see* Spectrum.
Spectrogram: a photograph of an astronomical spectrum.
Spectrograph: an instrument that spreads visible light or other electromagnetic radiation into a pattern, or spectrum, arranged by wavelength, and records the result photographically.
Spectrum: the array of electromagnetic radiation, arranged in order of wavelength, from long-wave radio to short-wave gamma rays. Also, a narrower band of wavelengths, as the visible spectrum, in which light dispersed by a prism or other means shows its component colors; often banded with absorption or emission lines.
Starburst: the sudden birth of many stars close together; the causes of this phenomenon are not yet well understood.
Supercomputer: a term applied to the fastest, most powerful computers of a given time. Supercomputers typically are used for complex simulations and scientific problems that involve large amounts of data.
Superconductivity: the propensity of certain materials to conduct electric current without resistance or loss of energy when greatly cooled.
Supergiant: a very massive, short-lived star.
Supernova: a stellar explosion that expels all or most of the star's mass and is extremely luminous.
Synchrotron emission: a type of nonthermal radiation generated by electrons and other charged particles spiraling around magnetic field lines at near light-speed.
Thermal radiation: electromagnetic energy produced by heat-related processes.
Transistor: a sealed device, consisting of grouped semiconductors, used to amplify or modify electric current and voltage.
T Tauri star: a very young star characterized by extensive and violent ejections of its mass.
Ultraviolet: a band of electromagnetic radiation with a higher frequency and shorter wavelength than visible blue light. Most ultraviolet is absorbed by the atmosphere, so ultraviolet astronomy is normally performed in space.
Van Allen belts: two zones of charged particles, trapped by Earth's magnetic field and concentrated at altitudes of about 3,000 and 10,000 miles. The particles come from the solar wind and from collisions of cosmic rays with atoms and molecules in Earth's atmosphere.
Very Large Array (VLA): a radio telescope, located in New Mexico, that consists of twenty-seven movable parabolic antennas arranged in the shape of a Y; each leg of the Y is about thirteen miles long. The VLA provides very high resolution for the study of radio sources.
Very long baseline interferometry (VLBI): the investiga-

tion of celestial radio sources through the simultaneous use of two or more radio telescopes separated by hundreds or thousands of miles. VLBI can employ a receiving array nearly as large as the Earth, producing very high resolution.

Wavelength: the distance from crest to crest or trough to trough of an electromagnetic or other wave. Wavelengths are related to frequency: The longer the wavelength, the lower the frequency.

White dwarf: an old, extremely dense star whose core has collapsed after exhausting its nuclear fuel of hydrogen and helium. A white dwarf with the mass of the Sun would be about the size of the Earth.

X-radiation (x-ray): radiation intermediate in wavelength between ultraviolet radiation and gamma rays. Because x-rays are absorbed by the atmosphere, x-ray astronomy is performed in space.

X-ray binary: a binary star system that is a source of x-rays, generated when matter from one star is accreted onto its more compact companion, which may be a white dwarf, neutron star, or black hole.

BIBLIOGRAPHY

Books

Abell, George O., David Morrison, and Sidney C. Wolff, *Exploration of the Universe* (5th ed.). New York: Saunders College Publishing, 1987.

Audouze, Jean, and Guy Israël, eds., *The Cambridge Atlas of Astronomy.* Cambridge, England: Cambridge University Press, 1983.

Bartusiak, Marcia, *Thursday's Universe.* New York: Times Books, 1986.

Bernstein, Jeremy, *Three Degrees above Zero.* New York: Charles Scribner's Sons, 1984.

Burbidge, Geoffrey, and A. Hewitt, eds., *Telescopes for the 1980s.* Palo Alto, Calif.: Annual Reviews, 1981.

Burbidge, Geoffrey, and Margaret Burbidge, *Quasi-Stellar Objects.* San Francisco, Calif.: W. H. Freeman, 1967.

Burbidge, Geoffrey, David Layzer, and John G. Phillips, eds.:
Annual Review of Astronomy and Astrophysics, Vol. 22. Palo Alto, Calif.: Annual Reviews, 1984.
Annual Review of Astronomy and Astrophysics, Vol. 25. Palo Alto, Calif.: Annual Reviews, 1987.

Clark, David H.:
The Cosmos from Space. New York: Crown, 1987.
The Quest for SS 433. New York: Penguin Books, 1985.

Cohen, Nathan, *Gravity's Lens: Views of the New Cosmology.* New York: John Wiley & Sons, 1988.

Computers and the Cosmos (Understanding Computers series). Alexandria, Va.: Time-Life Books, 1988.

Cornell, James, and John Carr, eds., *Infinite Vistas: New Tools for Astronomy.* New York: Charles Scribner's Sons, 1985.

Cornell, James, and Paul Gorenstein, eds., *Astronomy from Space: Sputnik to Space Telescope.* Cambridge, Mass.: MIT Press, 1983.

Cornell, James, and F. R. Harnden, Jr., *Visions of "Einstein."* Washington, D.C.: Office of Printing and Photographic Services, Smithsonian Institution, 1981.

Drever, Ronald P., et al., "Gravitational Wave Detectors Using Laser Interferometers and Optical Cavities: Ideas, Principles and Prospects." In *Quantum Optics, Experimental Gravity, and Measurement Theory,* ed. by P. Meystre and M. Scully. New York: Plenum, 1983.

Fernbach, S., and A. Taub, eds., *Computers and Their Role in the Physical Sciences.* New York: Gordon and Breach, 1970.

Field, George B., and Eric J. Chaisson, *The Invisible Universe.* Boston: Birkhäuser, 1985.

Harwit, Martin, *Cosmic Discovery: The Search, Scope, and Heritage of Astronomy.* New York: Basic Books, 1981.

Hawking, S. W., and W. Israel, eds., *Three Hundred Years of Gravitation.* Cambridge, England: Cambridge University Press, 1987.

Henbest, Nigel, ed., *Observing the Universe.* Oxford: Basil Blackwell, 1984.

Henbest, Nigel, and Michael Martin, *The New Astronomy.* Cambridge, England: Cambridge University Press, 1983.

Hill, Thomas W., and Richard A. Wolf, "Solar-Wind Interactions." In *The Upper Atmosphere and Magnetosphere.* Washington, D.C.: National Academy of Sciences, 1977.

Hirsh, Richard F., *Glimpsing an Invisible Universe.* Cambridge, England: Cambridge University Press, 1983.

Kaler, James B., *Stars and Their Spectra: An Introduction to the Spectral Sequence.* Cambridge, England: Cambridge University Press, 1989.

Kaufmann, William J., III, *Universe.* New York: W. H. Freeman, 1985.

Kondo, Y., ed., *Exploring the Universe with the IUE Satellite.* Dordrecht, Netherlands: D. Reidel, 1987.

Lamb, Donald Q., and Joseph Patterson, ed., *Cataclysmic Variables and Low-Mass X-Ray Binaries,* Dordrecht, Netherlands: D. Reidel, 1985.

Meyer, Peter, "Cosmic Radiation." In *Encyclopedia of Physical Science and Technology,* Vol. 3. New York: Academic Press, 1987.

Meystre, P., and M. Scully, *Quantum Optics, Experimental Gravity, and Measurement Theory.* New York: Plenum, 1983.

Ortega, James M., ed., *Computer Science and Scientific Computing.* New York: Academic Press, 1976.

Satterthwaite, Gilbert E., *Encyclopedia of Astronomy.* New York: St. Martin's Press, 1971.

Shapiro, Stuart L., and Saul A. Teukolsky, *Black Holes, White Dwarfs, and Neutron Stars.* New York: John Wiley & Sons, 1983.

Spitzer, Lyman, Jr., *Searching between the Stars.* New Haven, Conn.: Yale University Press, 1982.

Stalio, R., and L. A. Willson, eds., *Pulsation and Mass Loss in Stars.* Dordrecht, Netherlands: Kluwer Academic Publishers, 1988.

Tucker, Wallace, *The Star Splitters.* Washington, D.C.: NASA Scientific and Technical Information Branch, 1984.

Tucker, Wallace, and Riccardo Giacconi, *The X-Ray Universe.* Cambridge, Mass.: Harvard University

Press, 1985.
Tucker, Wallace, and Karen Tucker, *The Cosmic Inquirers.* Cambridge, Mass.: Harvard University Press, 1986.
Zeilik, Michael, *Astronomy: The Evolving Universe* (5th ed.). New York: John Wiley & Sons, 1988.

Periodicals

Armstrong, J. W., "Spacecraft Gravitational Wave Experiments." *Gravitational Wave Data Analysis,* 1989, pp. 153-172.
Armstrong, J. W., F. B. Estabrook, and H. D. Wahlquist, "A Search for Sinusoidal Gravitational Radiation in the Period Range 30-2000 Seconds." *Astrophysical Journal,* July 15, 1987.
"AstroNews: Was the Moon Formed by a Giant Collision?" *Astronomy,* July 1986.
Baltrusaitis, R. M., et al., "The Utah Fly's Eye Detector." *Nuclear Instruments and Methods in Physics Research,* 1985, pp. 410-428.
Barnes, Joshua, "Encounters of Disk/Halo Galaxies." *Astrophysical Journal,* August 15, 1988.
Barnes, Joshua, and Piet Hut, "A Hierarchical O(N log N) Force-Calculation Algorithm." *Nature,* December 4, 1986.
Bartusiak, Marcia, "Sensing the Ripples in Space-Time." *Science,* April 1985.
Beatty, J. Kelly, "The Making of a Better Moon." *Sky & Telescope,* December 1986.
Benz, Willy, Wayne L. Slattery, and A. G. W. Cameron:
"Collisional Stripping of Mercury's Mantle." *Icarus,* 1988, Vol. 74, pp. 516-528.
"The Origin of the Moon and the Single-Impact Hypothesis, I." *Icarus,* 1986, Vol. 66, pp. 515-535.
"The Origin of the Moon and the Single-Impact Hypothesis, II." *Icarus,* 1987, Vol. 71, pp. 30-45.
Bertout, Claude, Gibor Basri, and Jerome Bouvier, "Accretion Disks around T Tauri Stars." *Astrophysical Journal,* July 1, 1988.
Binns, W. R., et al., "Cosmic-Ray Energy Spectra between 10 and Several Hundred GeV Per Atomic Mass Unit for Elements from ${}_{18}$Ar to ${}_{28}$Ni: Results from *HEAO 3.*" *Astrophysical Journal,* January 15, 1988.
Broad, William J.:
"Soviet Space Radiation Hampering U.S. Satellite." *New York Times,* November 17, 1988.
"Space Pollution Forces NASA to Change Plans for Key Projects." *New York Times,* December 22, 1988.
Clark, George W., "New Instruments for Astronomy: High-Energy Astrophysics." *Physics Today,* November 1982.
Cowen, Ron, "Gravity Wave Sensors." *The World & I,* August 1987.
"Disrupting Galactic Jets." *Sky & Telescope,* June 1988.
Drever, Ronald P., "The Search for Gravitational Waves." *Engineering & Science,* January 1983.
Dwyer, Robert, and Peter Meyer, "Composition of Cosmic-Ray Nuclei from Boron to Nickel for 1200 to 2400 MeV Per Nucleon." *Astrophysical Journal,* July 15, 1989.
Eakin, Julie Sinclair, "Tuscon Takes on Sky Glow." *Light,* fall 1986.
Engelmann, J. J., et al. "Source Energy Spectra of Heavy Cosmic Ray Nuclei as Derived from the French-Danish Experiment on HEAO-3." *Astronomy and Astrophysics,* 1985, Vol. 148, pp. 12-20.
Fisher, Arthur, "Science Newsfront: Birth of a Quasar." *Popular Science,* September 1987.
Giacconi, Riccardo, et al., "Evidence for X Rays from Sources Outside the Solar System." *Physical Review Letters,* December 1, 1962.
Goulding, Fred S., "Search for the Neutrino: Tracking the Elusive Particle for Clues to the Missing Mass of the Universe." *LBL Research Review,* summer 1985.
Grunsfeld, John M., et al. "Energy Spectra of Cosmic-Ray Nuclei from 50 to 2000 GeV per AMU." *Astrophysical Journal,* April 1, 1988.
Hartley, Karen, "Mixing It Up in Space." *Science News,* April 8, 1989.
Helfand, David:
"Bang: The Supernova of 1987." *Physics Today,* August 1987.
"Fleet Messengers from the Cosmos." *Sky & Telescope,* March 1988.
"Hello Darkness, My Old Friend . . ." *News! from the Naval Observatory,* July 1988.
Henbest, Nigel, "Satellites Signal the End for Radio Astronomy." *New Scientist,* December 3, 1988.
Hendry, Allan, "Light Pollution: A Status Report." *Sky & Telescope,* June 1984.
Hirata, K. S., et al. "Observation in the Kamiokande-II Detector of the Neutrino Burst from Supernova SN1987A." *Physical Review D: Particles and Fields,* July 15, 1988.
Hjellming, R. M., and K. J. Johnston, "An Analysis of the Proper Motions of SS 433 Radio Jets." *Astrophysical Journal,* June 15, 1981.
Holmberg, Erik, "On the Clustering Tendencies among the Nebulae." *Astrophysical Journal,* November 1941.
Hones, Edward W., Jr.:
"The Earth's Magnetotail." *Scientific American,* March 1986.
"Transient Phenomena in the Magnetotail and Their Relation to Substorms." *Space Science Reviews,* 1979, Vol. 23, pp. 393-410.
Hones, Edward W., and P. R. Higbie, "Distribution and Detection of Positrons from an Orbiting Nuclear Reactor." *Science,* April 28, 1989.
Hunter, Tim B., and Bob Goff, "Shielding the Night Sky." *Astronomy,* September 1988.
Hut, Piet, and Gerald Jay Sussman, "Advanced Computing for Science." *Scientific American,* October 1987.
"IUE: A Miraculous Recovery." *Sky & Telescope,* February 1986.
"IUE: Ten Years and Still Working." *Science News,* February 6, 1988.
Jeffries, Andrew D., et al., "Gravitational Wave Observatories." *Scientific American,* June 1987.
Kaler, James B., "Planetary Nebulae and the Death of Stars." *American Scientist,* May-June 1986.
Kanipe, Jeff, "Anatomy of a Cosmic Jet." *Astronomy,* July 1988.
Keel, William C., "Crashing Galaxies, Cosmic Fireworks." *Sky & Telescope,* January 1989.
Kellerman, K. I., and A. R. Thompson, "The Very Long Baseline Array." *Science,* July 12, 1985.
Koshiba, Masa-Toshi, "Observational Neutrino Astrophysics." *Physics Today,* December 1987.

Kunzig, Robert, "Iron Planet." *Discover,* February 1989.
Lamb, R. C., and T. C. Weekes, "Very High Energy Gamma-Ray Binary Stars." *Science,* December 11, 1987.
Lindley, David, "Keeping Astronomers in the Dark." *Nature,* January 21, 1988.
Maddox, John, "New Ways with Gravity Waves." *Nature,* December 8, 1988.
Maran, Stephen P., "Deterioration of the Skies Alarms Astronomers." *IAU Today,* August 9, 1988.
Margon, Bruce, "The Bizarre Spectrum of SS 433." *Scientific American,* October 1980.
Melott, Adrian L., and Sergei F. Shandarin, "Gravitational Instability with High Resolution." *Astrophysical Journal,* August 1988.
Napier, Peter J., A. Richard Thompson, and Ronald D. Ekers, "The Very Large Array: Design and Performance of a Modern Synthesis Radio Telescope." *Proceedings of the IEEE,* November 11, 1983.
Norman, Michael L., Jack O. Burns, and Martin E. Sulkanen, "Disruption of Galactic Radio Jets by Shocks in the Ambient Medium." *Nature,* September 8, 1988.
"Observe with the HST!" *Sky & Telescope,* February 1986.
O'Neill, Terrence J., et al., "Observations of Nuclear Reactors on Satellites with a Balloon-Borne Gamma-Ray Telescope." *Science,* April 28, 1989.
Overbye, Dennis, "Does Anyone Understand SS 433?" *Sky & Telescope,* December 1979.
Pankonin, Vernon, and R. Marcus Price, "Radio Astronomy and Spectrum Management: The Impact of WARC-79." *IEEE Transactions on Electromagnetic Compatibility,* August 1981.
Passavant, Tom, "Seeing the Light." *Light,* fall 1986.
Peterson, Ivars, "Relativity by the Numbers." *Science News,* September 1988.
Readhead, Anthony C. S., "Radio Astronomy by Very-Long-Baseline Interferometry." *Scientific American,* June 1982.
Rieger, Erich, et al., "Man-Made Transients Observed by the Gamma-Ray Spectrometer on the Solar Maximum Mission Satellite." *Science,* April 28, 1989.
Schorn, Ronald A., "Neutrinos from Hell." *Sky & Telescope,* May 1987.
Schroeder, Michael C., and Neil F. Comins, "Galactic Collisions on Your Computer." *Astronomy,* December 1988.
Schulman, Lawrence S., and Philip E. Seiden, "Percolation and Galaxies." *Science,* July 25, 1986.
Schweizer, François, "Colliding and Merging Galaxies." *Science,* January 17, 1986.
"Selling the Night Sky . . . ." *News! from the Naval Observatory,* September 1988.
Shapiro, S. L., and S. A. Teukolsky, "Building Black Holes: Supercomputer Cinema." *Science,* July 22, 1988.
Share, G. H., et al., "Geomagnetic Origin for Transient Particle Events from Nuclear Reactor-Powered Satellites." *Science,* April 28, 1989.
Shore, Lys Ann, "The Telescope That Never Sleeps." *Astronomy,* August 1987.
Sullivan, Woodruff T., III:
"Our Endangered Night Skies." *Sky & Telescope,* May 1984.
"A 10km Resolution Image of the Entire Night-Time Earth Based on Cloud-Free Satellite Photographs in the 400-1100nm Band." *International Journal of Remote Sensing,* 1989, Vol. 10, No. 1, pp. 1-5.
"Taking Back the Night Sky." *Science,* September 9, 1988.
Talcott, Richard, "Insight into Star Death." *Astronomy,* February 1988.
Thomsen, Dietrick E., "A Satellite Triangle for Gravity Waves." *Science News,* January 3, 1987.
Tierney, John, "Exploding Star Contains Atoms of Elvis Presley's Brain." *Discover,* July 1987.
Toomre, Alar, and Juri Toomre:
"Galactic Bridges and Tails." *Astrophysical Journal,* December 15, 1972.
"Violent Tides between Galaxies." *Scientific American,* December 1973.
Trimble, Virginia, "Gravity Waves: A Progress Report." *Sky & Telescope,* October 1987.
Turner, Edwin L., "Gravitational Lenses." *Scientific American,* July 1988.
Waldrop, M. Mitchell, "Supernova Neutrinos at IMB." *Science,* March 13, 1987.
Watson, M. G., et al., "The X-Ray Lobes of SS 433." *Astrophysical Journal,* October 15, 1983.

Other Sources

California Institute of Technology and the Massachusetts Institute of Technology, "Laser Interferometer Gravitational Wave Observatory." Unpublished brief, February 3, 1986.
"High Energy Astronomy Observatory." Washington, D.C.: Public Affairs Division, NASA Headquarters, no date.
Jones, C., C. Stern, and W. Forman, "New Windows on the Universe: The NASA Great Observatories." Slide and text presentation. Cambridge, Mass.: Harvard-Smithsonian Center for Astrophysics, 1987.
"Orbiting Nuclear Reactors Disrupt Civilian Satellite Missions." Press release. Los Angeles: Committee to Bridge the Gap, September 1988.
Reetz, Arthur J., "Transient Gamma-Ray Events." NASA memorandum, Washington, D.C., August 1988.

INDEX

Numerals in italics indicate an illustration of the subject mentioned.

A

ACKNOWLEDGMENTS

The editors wish to thank Tony Acevedo, Arecibo Observatory, Arecibo, Puerto Rico; John Armstrong, Pasadena, Calif.; Gibor Basri, University of California, Berkeley; Steven V. W. Beckwith, Cornell University, Ithaca, N.Y.; Willy Benz, Harvard College Observatory, Cambridge, Mass.; G. Dana Berry, Space Telescope Science Institute, Baltimore; Dieter Brill, University of Maryland, College Park; Dan Brocius, Whipple Observatory, Amado, Ariz.; Eric Chaisson, Space Telescope Science Institute, Baltimore; Geoff Chester, National Air and Space Museum, Washington, D.C.; Tom Clark, Tom Cline, NASA Goddard Space Flight Center, Greenbelt, Md.; Bill Dent, Royal Observatory, Edinburgh, Scotland; John Dickel, University of Illinois, Urbana; Martin Elvis, Harvard-Smithsonian Center for Astrophysics, Cambridge, Mass.; Guisseppina Fabbiano, Harvard-Smithsonian Center for Astrophysics, Cambridge, Mass.; Shing Fung, NASA Goddard Space Flight Center, Greenbelt, Md.; Owen Gingerich, Harvard-Smithsonian Center for Astrophysics, Cambridge, Mass.; Bob Havlen, Very Large Array, National Radio Astronomy Observatory, Socorro, N.Mex.; David Helfand, Columbia University, New York; Lars Hernquist, Institute for Advanced Studies, Princeton, N.J.; Eugen Hintsches, Max Planck Gesellschaft, Munich; Erik Holmberg, Partille, Sweden; Garth Hull, NASA Ames Research Center, Mountain View, Calif.; Icko Iben, Jr., University of Illinois, Urbana; Ken Johnston, James Kurfess, Naval Research Laboratory, Washington, D.C.; Donald Lamb, Jacques L'Heureaux, University of Chicago; Carol J. Lonsdale, Infrared Processing and Analysis Center, California Institute of Technology, Pasadena; Al Mann, University of Pennsylvania, Philadelphia; Bruce Margon, University of Washington, Seattle; Ian McLean, Joint Astronomy Center, Hilo, Hawaii; Adrian Melott, University of Kansas, Lawrence; Peter Meyer, Enrico Fermi Institute—LASR, University of Chicago; Joyce Milner, Wallops Flight Facility, Wallops Island, Va.; Warren Moos, Johns Hopkins University, Baltimore; Richard Mushotsky, NASA Goddard Space Flight Center, Greenbelt, Md.; Roger Norrod, National Radio Astronomy Observatory, Greenbank, W.Va.; Sten Odenwald, Naval Research Laboratory, Washington, D.C.; John Payne, National Radio Astronomy Observatories, Charlottesville, Va.; Read Predmore, University of Massachusetts, Amherst; Rolf Schwarz, Max Planck Institute, Bonn, West Germany; François Schweizer, Department of Terrestrial Magnetism, Washington, D.C.; P. K. Seidelmann, Naval Observatory, Washington, D.C.; Philip Seiden, IBM Watson Research Center, Yorktown Heights, N.Y.; Fred Seward, CIA, Cambridge, Mass.; Stuart Shapiro, Cornell University, Ithaca, N.Y.; Patricia Smiley, National Radio Astronomy Observatories, Charlottesville, Va.; Theodore Snow, University of Colorado, Boulder; George Sonneborn, Ted Stecher, NASA Goddard Space Flight Center, Greenbelt, Md.; Stephen Strom, University of Massachusetts, Amherst; Joseph Taylor, Princeton University, Princeton, N.J.; Saul Teukolsky, Cornell University, Ithaca, N.Y.; Edwin Turner, Princeton University, Princeton, N.J.; Gerrit L. Verschuur, Bowie, Md.; Ray Villard, Space Telescope Science Institute, Baltimore; Dr. R. Vogt, California Institute of Technology, Pasadena; Fred Whipple, Cambridge, Mass.; Richard Wielebinski, Thomas Wilson, Max Planck Institute, Bonn, West Germany.

PICTURE CREDITS

The sources for the illustrations in this book are listed below. Credits from left to right are separated by semicolons; credits from top to bottom are separated by dashes.

Cover: Art by Stephen R. Wagner. Front and back endpapers: Artwork by Time-Life Books. 2: Courtesy Farhad Yusef-Zadeh, Northwestern University. 3: Mark McCaughrean/NASA Goddard Space Flight Center. 4, 5: David Malin/Anglo-Australian Telescope Board. 6: SPL, London, courtesy Ralph C. Bohlin, Space Telescope Science Institute, and Theodore P. Stecher, NASA Goddard Space Flight Center. 7: Harvard-Smithsonian Center for Astrophysics. 12, 13: John R. Dickel, Stephen S. Murray, Jeffrey Morris, and Donald Wells, University of Illinois. 14: Initial cap, detail from pages 12, 13. 16: Courtesy NRAO/AUI. 17: SPL, London, courtesy Ralph C. Bohlin, Space Telescope Science Institute, and Theodore P. Stecher, NASA Goddard Space Flight Center. 18, 19: Astronomical Society of the Pacific; Popperfoto, London; Josh Young, University of Arizona; courtesy AT & T Archives; Robert P. Matthews, Princeton University. Background art by Stephen R. Wagner. 22, 23: W. T. Sullivan III. Background art by Mark Robinson. 24, 25: D. Crawford and B. Robinson, NOAO (2); U.K. Schmidt Telescope Unit. Art by Mark Robinson. 26-29: Line art for chart by Time-Life Books, based on material provided by Office of Spectrum Management, U.S. Department of Commerce. Background art by Mark Robinson. 31: The Arecibo Observatory is part of the National Astronomy and Ionosphere Center, which is operated by Cornell University under contract with the National Science Foundation. 33: James Sugar/Black Star. 34-37: Art by Stephen Bauer. 39: Courtesy NRAO/AUI. 40: Steven Beckwith and Anneila Sargent, Cornell University. 42, 43: NASA Ames Research Center. 44, 45: JPL/NASA. 47: NASA no. G-77-05354, courtesy Sky Publishing Corp. 49-59: Art by Joe Bergeron. 60, 61: Courtesy Fred Seward, Harvard-Smithsonian Center for Astrophysics. Inset courtesy NOAO. 62: Initial cap, detail from pages 60, 61. 64: Art by Sam Ward. 66, 67: Art by Stephen Bauer; photo courtesy NRAO/AUI. 68, 69: Art by Stephen Bauer; photos courtesy NRAO/AUI—NOAO. 70, 71: F. R. Harnden, Jr., Harvard-Smithsonian Center for Astrophysics. Line art by Time-Life Books. 72: Kent Wood, Naval Research Lab. 73: Artwork by Time-Life Books. 74, 75: NASA, Washington, D.C. 76, 77: Artwork by Time-Life Books—European Space Agency, *COS-B* Project Scientist, Kevin Bennett. 78, 79: Giovanni Fazio, Harvard-Smithsonian Center for Astrophysics. 80-85: Art by Alfred Kamajian. 86: Gary Ladd. 88: G. L. Cassiday, University of Utah. 91: Art by Sam Ward. 93: Art by Stephen R. Wagner. 94, 95: Art by Stephen R. Wagner—photo courtesy Prof. Richard Wielebinski and Dr. Wolfgang Reich, Max Planck Institut für Radioastronomie, Bonn. 96, 97: Art by Stephen R. Wagner; photo courtesy Thomas R. Geballe/UKIRT. 98, 99: Art by Stephen R. Wagner; photos courtesy John MacKenty, Don Schneider, Space Telescope Science Institute (2). 100, 101: Art by Stephen R. Wagner—photo courtesy Oswald Siegmund, University of California, Berkeley. 102, 103: Photo courtesy Smithsonian Astrophysical Observatory—art by Stephen R. Wagner. 104, 105: Photo courtesy European Space Agency, *COS-B* Project Scientist, Kevin Bennett—art by Stephen R. Wagner. 106, 107: Adrian Melott, University of Kansas. 108: Initial cap, detail from pages 106, 107. 110-115: Art by Damon M. Hertig. 118, 119: Art by Damon M. Hertig—photos courtesy Sanjoy Ghosh/NASA Goddard Space Flight Center (3). 122: Cornell University photograph by Charles Harrington. 125: Art by Sam Ward—photos courtesy Norman, Cox/University of Illinois, NCSA, and Burns, Sulkanen/University of New Mexico (3). 127-133: Background art and panels by Time-Life Books. 127-129: Photos courtesy Lars Hernquist and Joshua Barnes. 130, 131: Photos courtesy W. Benz, W. L. Stanley, and A. G. W. Cameron (5)—Stuart L. Shapiro and Saul Teukolsky, Cornell University (5). 132, 133: Photos courtesy Lars Hernquist.

Time-Life Books Inc.
is a wholly owned subsidiary of
THE TIME INC. BOOK COMPANY

President and Chief Executive Officer:
Kelso F. Sutton
President, Time Inc. Books Direct:
Christopher T. Linen

TIME-LIFE BOOKS INC.

EDITOR: George Constable
Executive Editor: Ellen Phillips
Director of Design: Louis Klein
Director of Editorial Resources: Phyllis K. Wise
Editorial Board: Russell B. Adams, Jr., Dale M. Brown, Roberta Conlan, Thomas H. Flaherty, Lee Hassig, Jim Hicks, Donia Ann Steele, Rosalind Stubenberg
Director of Photography and Research:
John Conrad Weiser

PRESIDENT: John M. Fahey, Jr.
Senior Vice Presidents: Robert M. DeSena, James L. Mercer, Paul R. Stewart, Curtis G. Viebranz, Joseph J. Ward
Vice Presidents: Stephen L. Bair, Stephen L. Goldstein, Marla Hoskins, Juanita T. James, Andrew P. Kaplan, Susan J. Maruyama, Robert H. Smith
Supervisor of Quality Control: James King

PUBLISHER: Joseph J. Ward

Editorial Operations
Copy Chief: Diane Ullius
Production: Celia Beattie
Library: Louise D. Forstall

Correspondents: Elisabeth Kraemer-Singh (Bonn), Christina Lieberman (New York), Maria Vincenza Aloisi (Paris), Ann Natanson (Rome). Valuable assistance was also provided by Lesley Coleman, Christine Hinze (London), and Mary Johnson (Stockholm).

VOYAGE THROUGH THE UNIVERSE

SERIES DIRECTOR: Roberta Conlan
Series Administrator: Judith W. Shanks

Editorial Staff for *The New Astronomy*
Designer: Ellen Robling
Associate Editors: Kristin Baker Hanneman, Blaine Marshall (pictures)
Text Editors: Peter Pocock (principal), Pat Daniels, Allan Fallow
Researchers: Karin Kinney, Mary McCarthy, Barbara Sause
Writer: Robert M. S. Somerville
Assistant Designer: Barbara M. Sheppard
Copy Coordinator: Darcie Conner Johnston
Picture Coordinator: Ruth Moss
Editorial Assistants: Jayne A. L. Dover, Katie Mahaffey

Special Contributors: Sarah Brash, Ken Crosswell, Jim Dawson, Alan MacRobert, Gina Maranto, Jim Merritt, Dennis Overbye, Dan Stashower, John Sullivan, Karen Tucker, and Wallace Tucker (text); Sydney Baily, Vilasini Balakrishnan, Patti Cass, Andrea Correll, Adam Dennis, Ed Dixon, Mark Galan, Sydney Johnson, Jocelyn Lindsay, Valerie May, Hugh McIntosh, Elizabeth Thompson (research); Barbara L. Klein (index).

CONSULTANTS

JOSHUA E. BARNES, an astrophysicist at the Institute for Advanced Studies, Princeton, New Jersey, specializes in computer modeling.

ROBERT A. BROWN is an astronomer with the Space Telescope Science Institute, Baltimore, Maryland.

CARL FICHTEL is a gamma ray scientist at NASA Goddard Space Flight Center in Greenbelt, Maryland.

ED FOMALONT specializes in radio interferometry at the National Radio Astronomy Observatories, Charlottesville, Virginia.

TOM GERGELY is in charge of interference protection of radio observatories at the National Science Foundation, Washington, D.C.

SANJOY GHOSH, a space physicist at NASA Goddard Space Flight Center, does computer modeling of astrophysical phenomena.

PAUL GORENSTEIN specializes in x-ray astronomy at the Harvard-Smithsonian Center for Astrophysics, Cambridge, Massachusetts.

RICHARD GRIFFITHS is an optical astronomer at the Space Telescope Science Institute.

JAMES B. KALER teaches stellar astronomy at the University of Illinois, Urbana-Champaign.

MARK McCAUGHREAN is an infrared Astronomer at NASA Goddard Space Flight Center.

STEVE MARAN, a senior staff scientist at NASA Goddard Space Flight Center, writes widely in astronomy and is press officer for the American Astronomical Society.

MICHAEL L. NORMAN is a research scientist at the National Center for Supercomputing Applications at the University of Illinois, Urbana-Champaign.

HARRY SHIPMAN, author of a number of astronomy books, teaches astronomy at the University of Delaware, Newark.

OSWALD SIEGMUND, a research physicist at the University of California, Berkeley, develops ultraviolet detector systems for satellites.

ALAR TOOMRE teaches mathematics at the Massachusetts Institute of Technology. He and his brother Juri were among the first to use computer simulations to study colliding galaxies.

JACK TUELLER, an astrophysicist at NASA Goddard Space Flight Center, specializes in the development of experimental sensing techniques, including high resolution gamma ray spectroscopy.

T. L. WILSON is a radio astronomer at the Max Planck Institute, Bonn, West Germany.

Library of Congress Cataloging in Publication Data
The New astronomy/by the editors of Time-Life Books.
p. cm. (Voyage through the universe).
Bibliography: p.
Includes index.
ISBN 0-8094-6883-2.
ISBN 0-8094-6884-0 (lib. bdg.).
1. Astronomy—Popular works. I. Time-Life Books.
II. Series.
QB44.2.N48 1989
520—dc20 89-4465 CIP

For information on and a full description of any of the Time-Life Books series, please call 1-800-621-7026 or write:
Reader Information
Time-Life Customer Service
P.O. Box C-32068
Richmond, Virginia 23261-2068

First printing. Printed in U.S.A.
Published simultaneously in Canada.
School and library distribution by Silver Burdett Company, Morristown, New Jersey 07960.

Earth: diameter 7,926 miles
Neptune: diameter 30,200 miles
Uranus: diameter 31,600 miles
Red supergiant: diameter 400 million miles
Solar System: diameter 7.5 billion miles
Globular cluster: diameter 2×10^{14} miles
Milky Way: diameter 100,000 light-years
Local Group of galaxies:
6 million light-years across
Largest double radio source:
length 17 million light-years